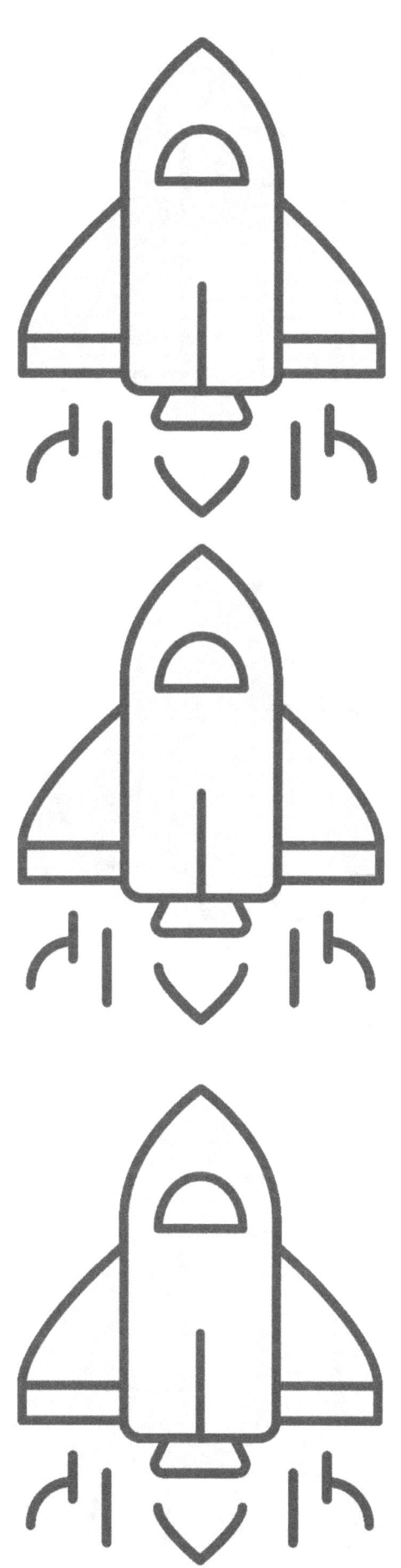
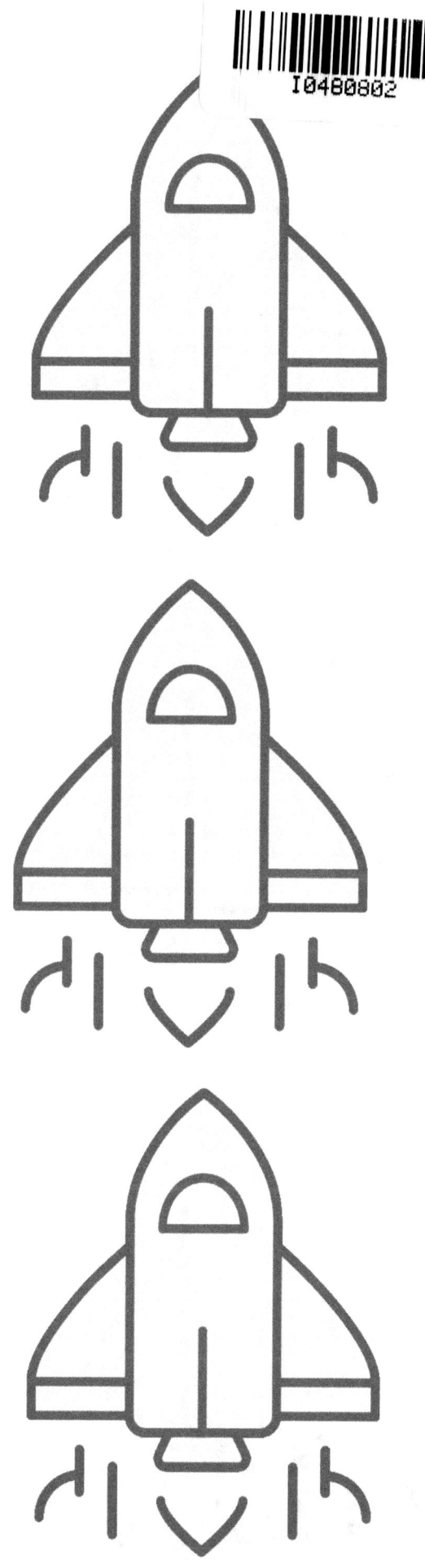

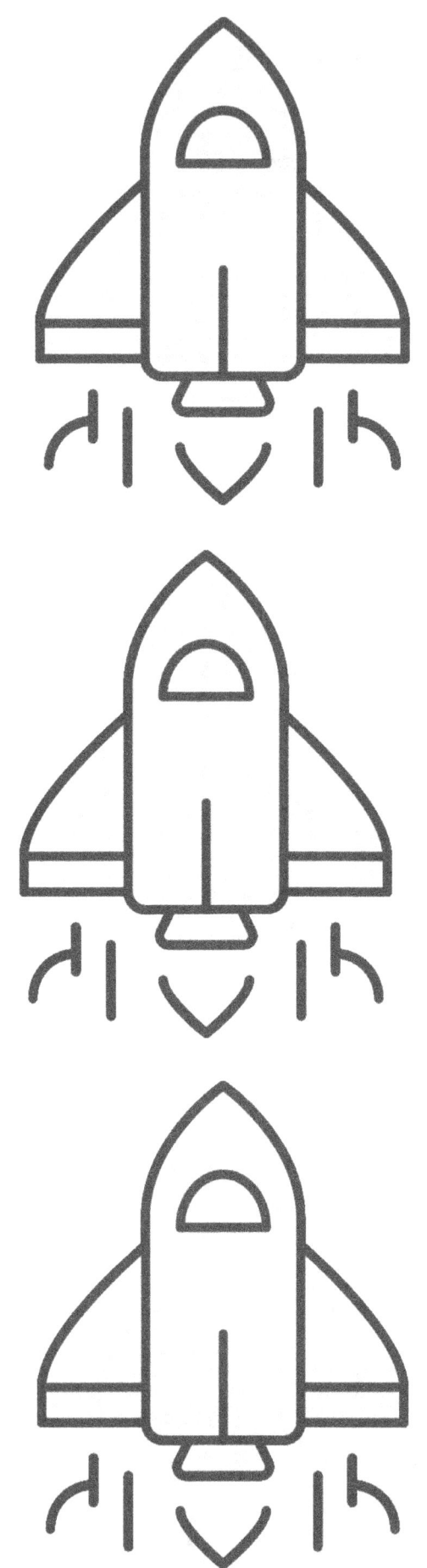

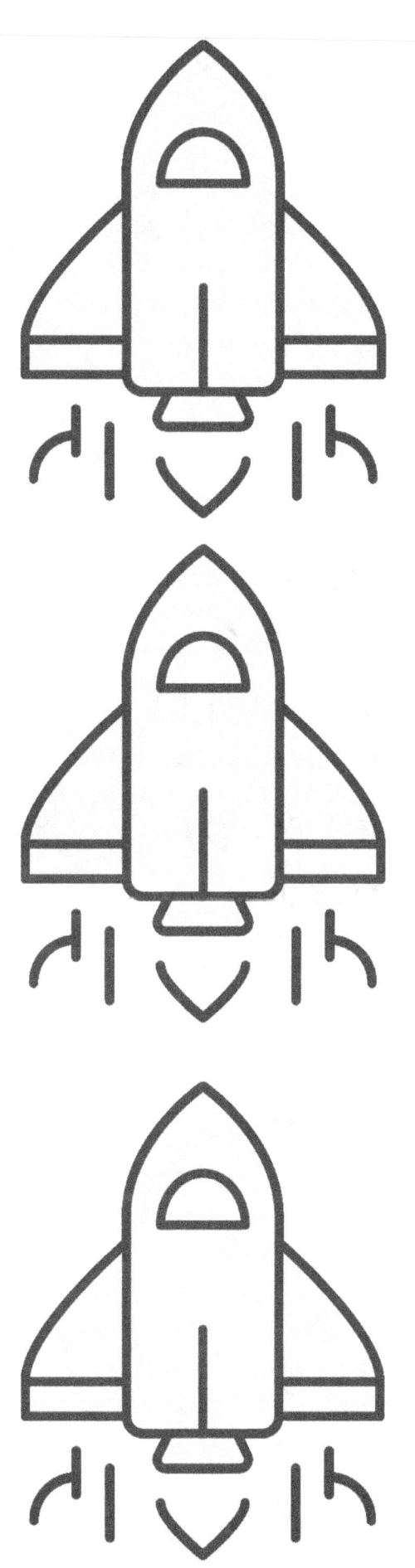
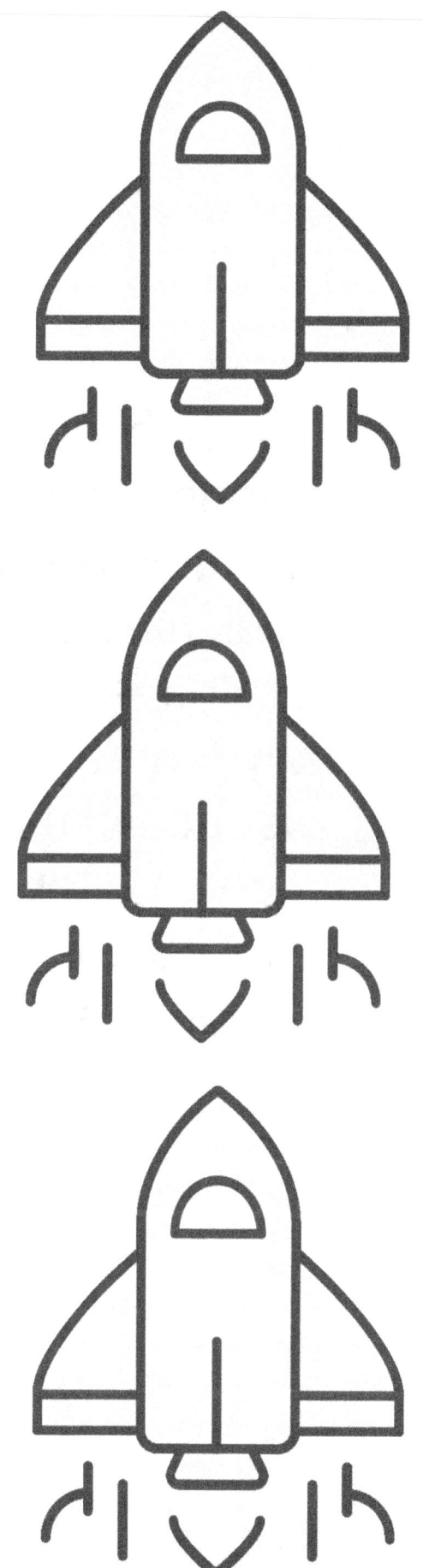

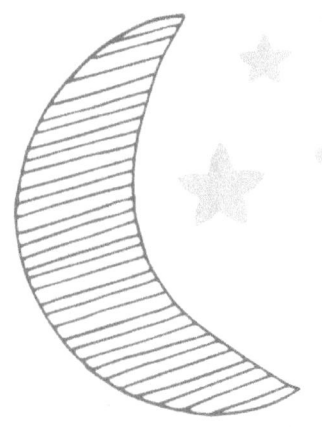

DOODLE

ME!

SPACE

Rhonda Ragan
Shuck

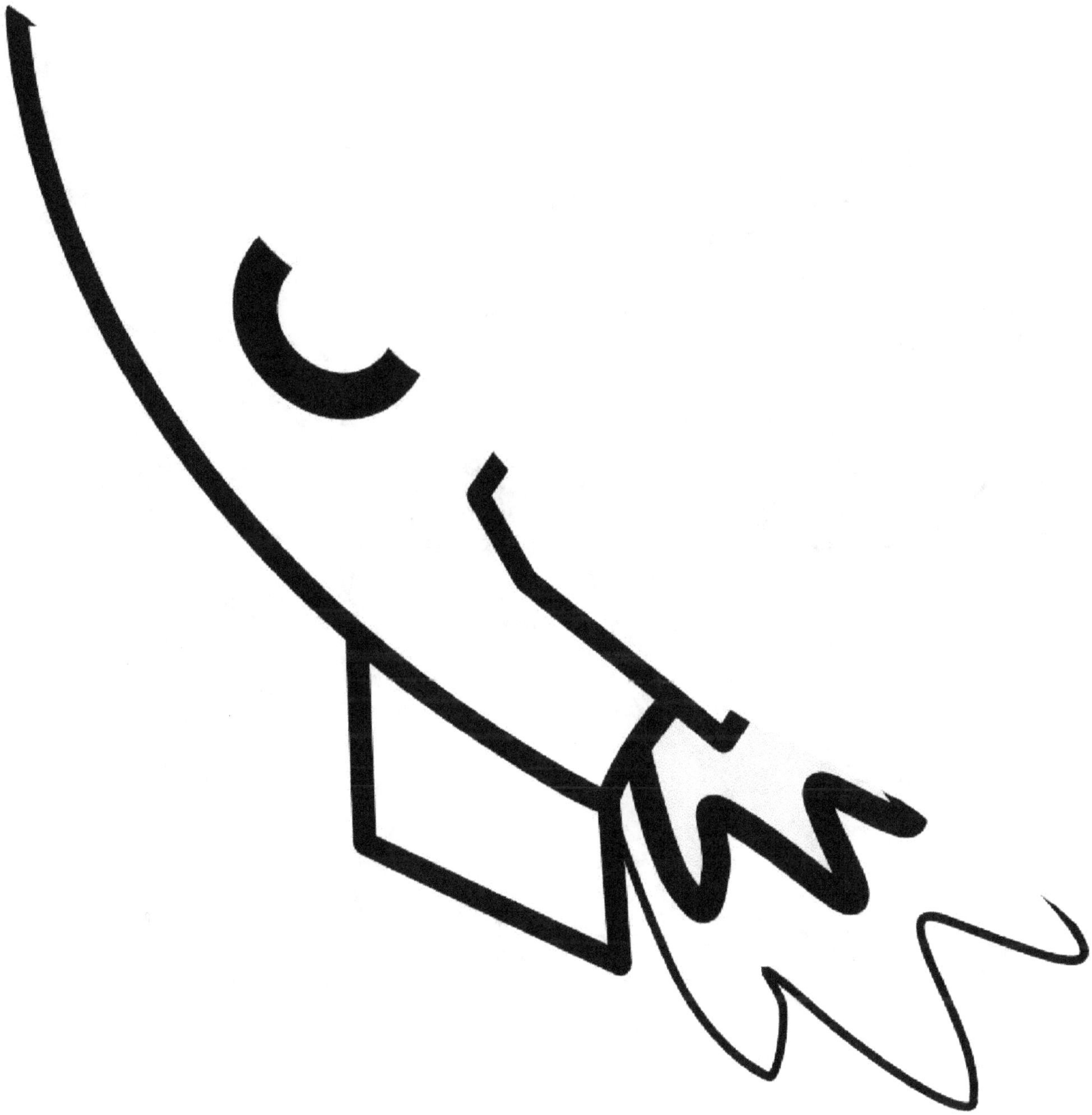

Draw the moon and stars.

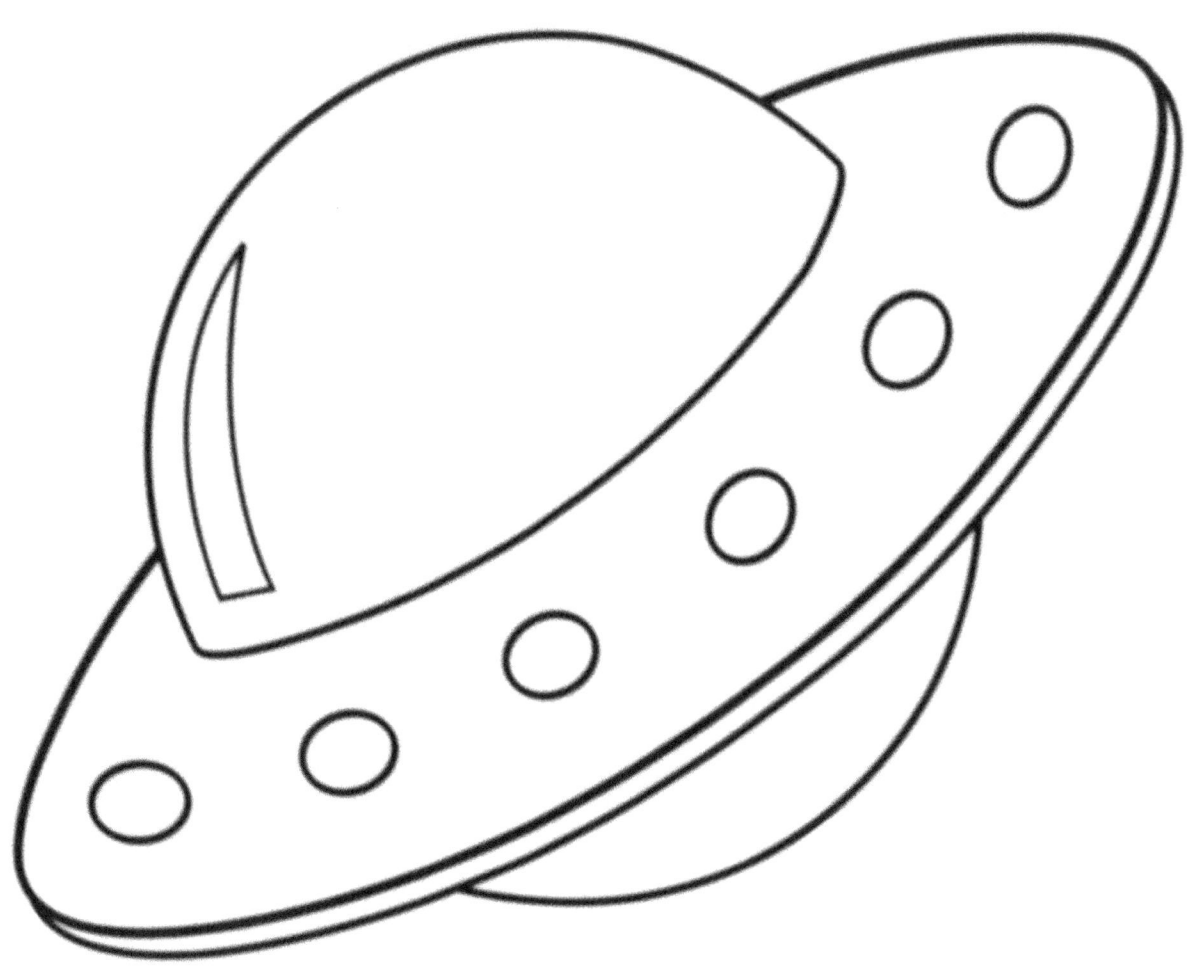

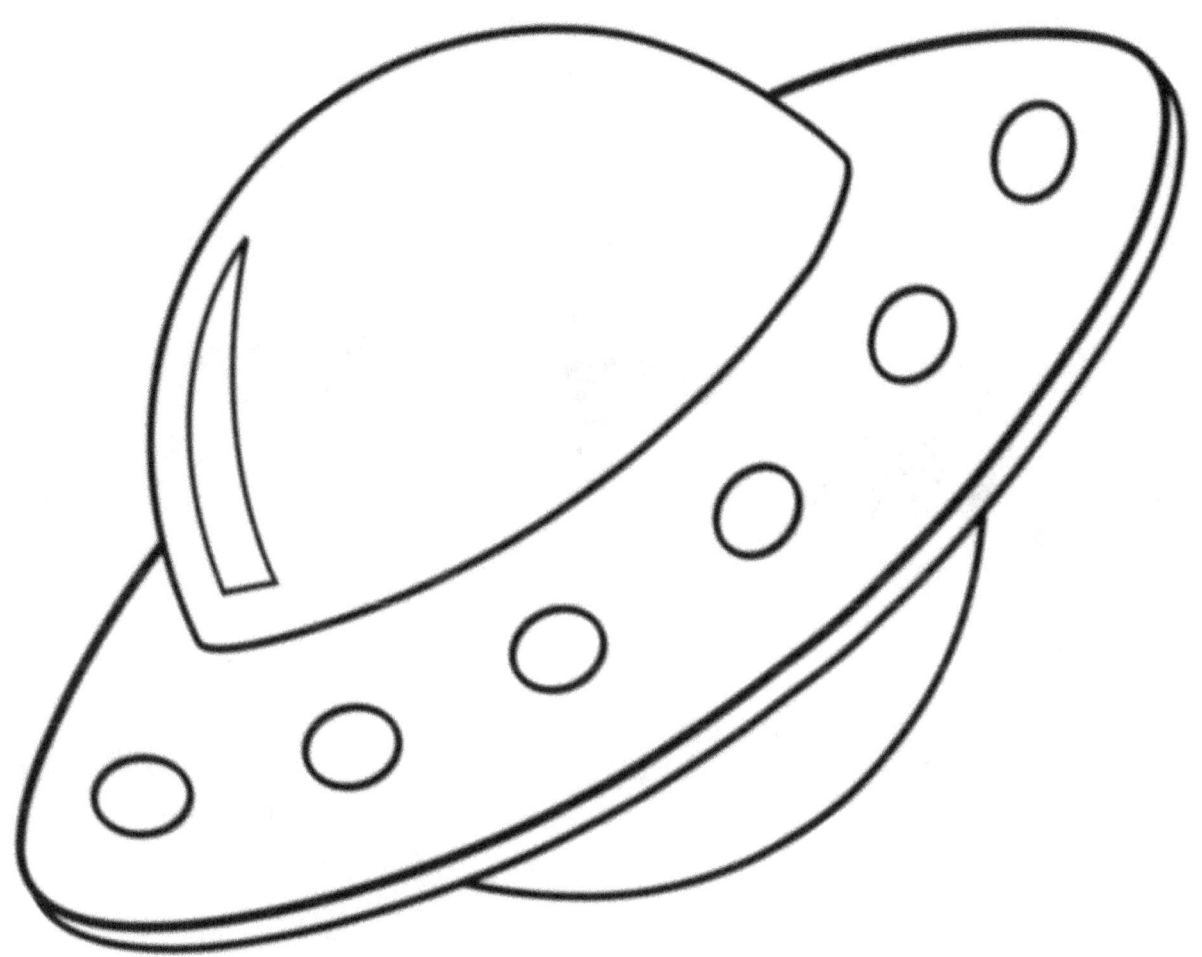

Draw an alien driving
the flying ship.

Count down to
blast off..... 10, 9, 8

7, 6, 5, 4, 3, 2 ,1
BLAST OFF!

Draw in the faces.

Can you draw
yourself flying in
space?

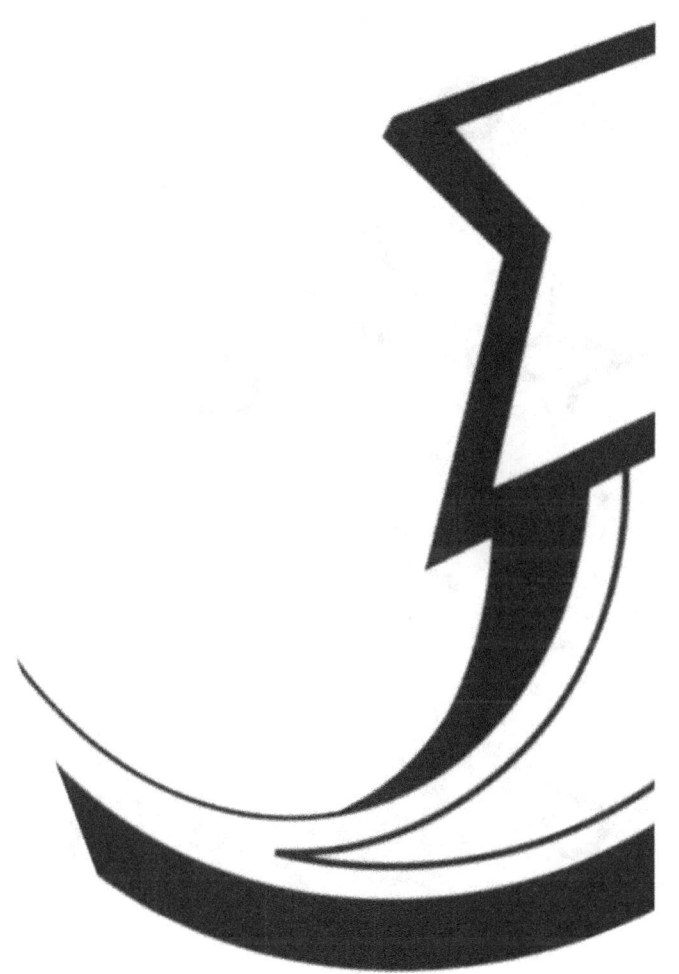

Draw you own set of stars.

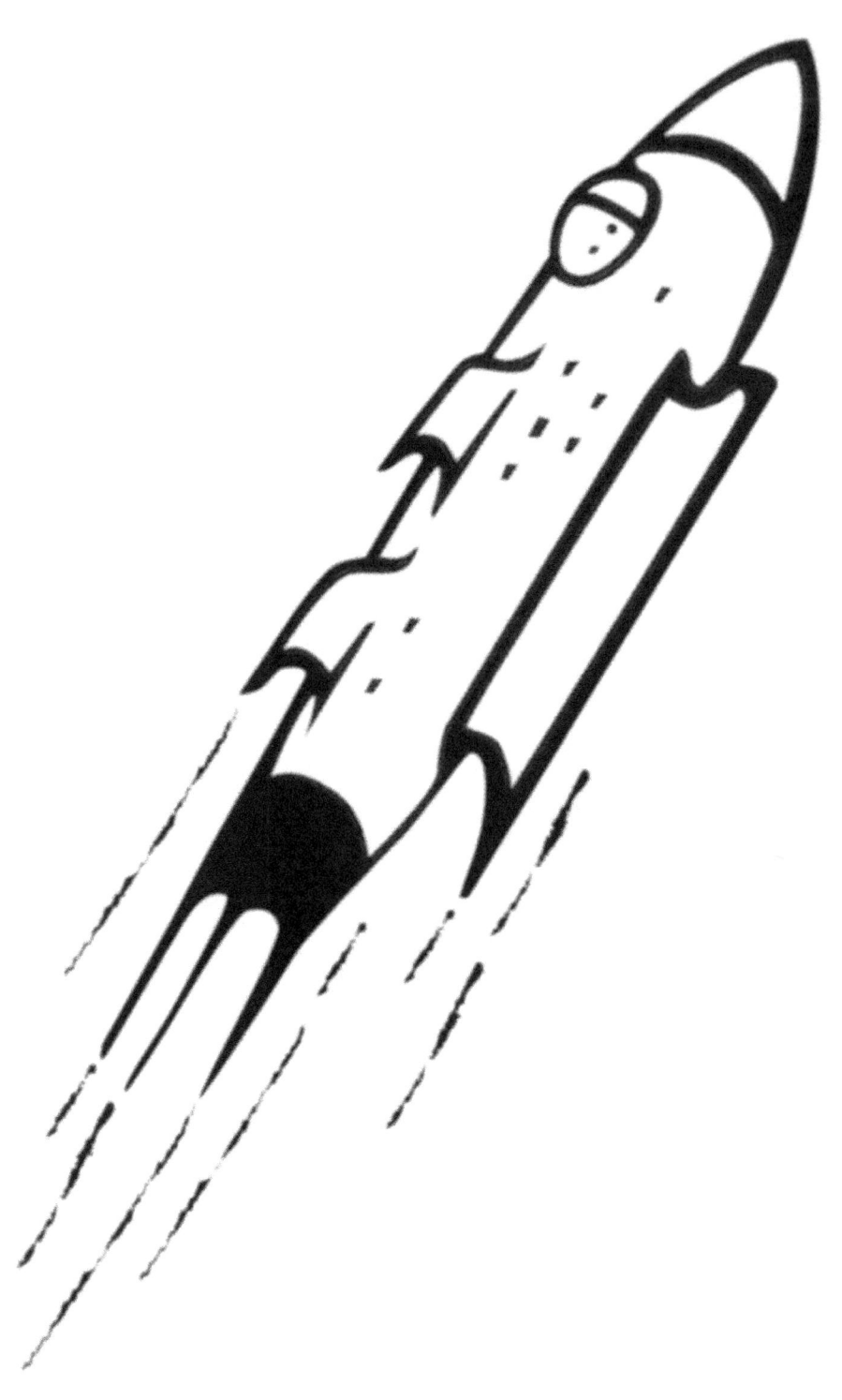

Draw a rocket going
to the moon.

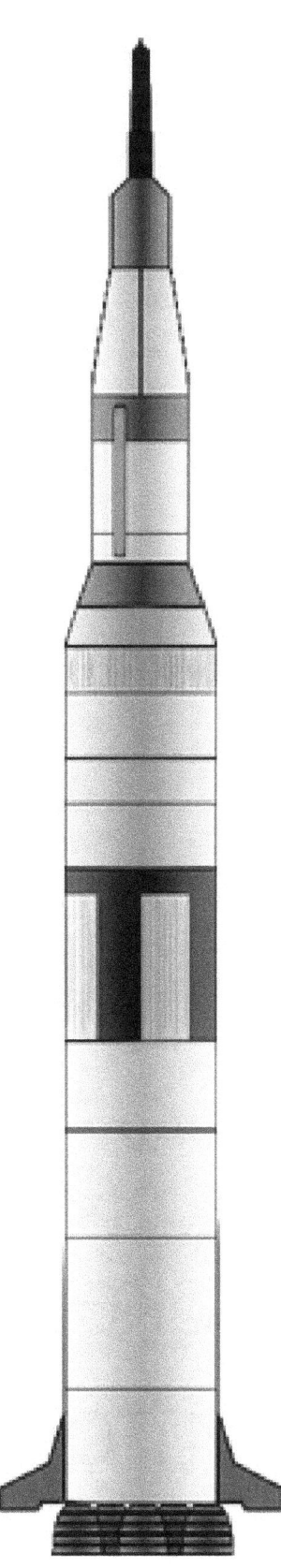

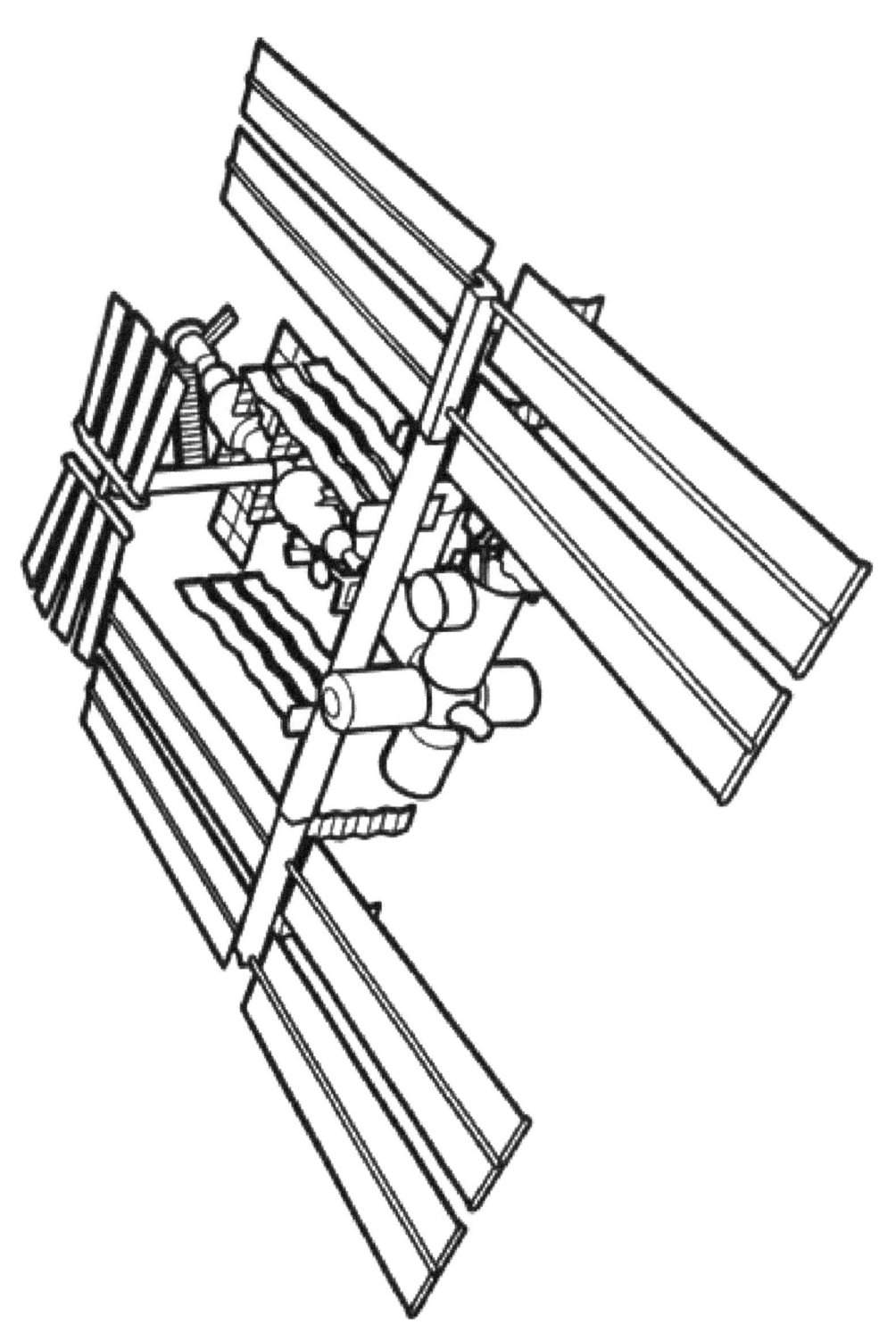

Design a Skylab.

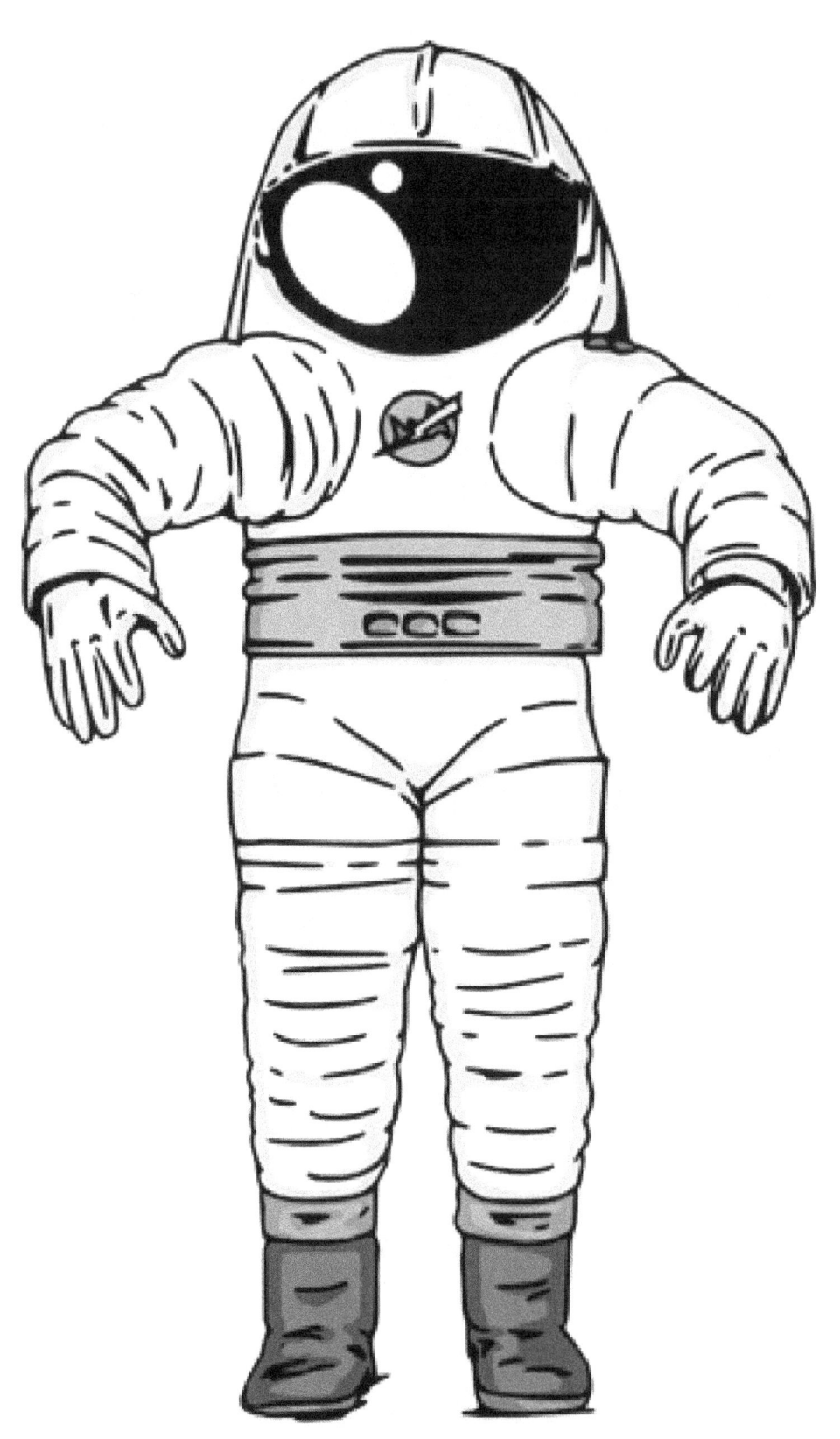

Design your space
suit.

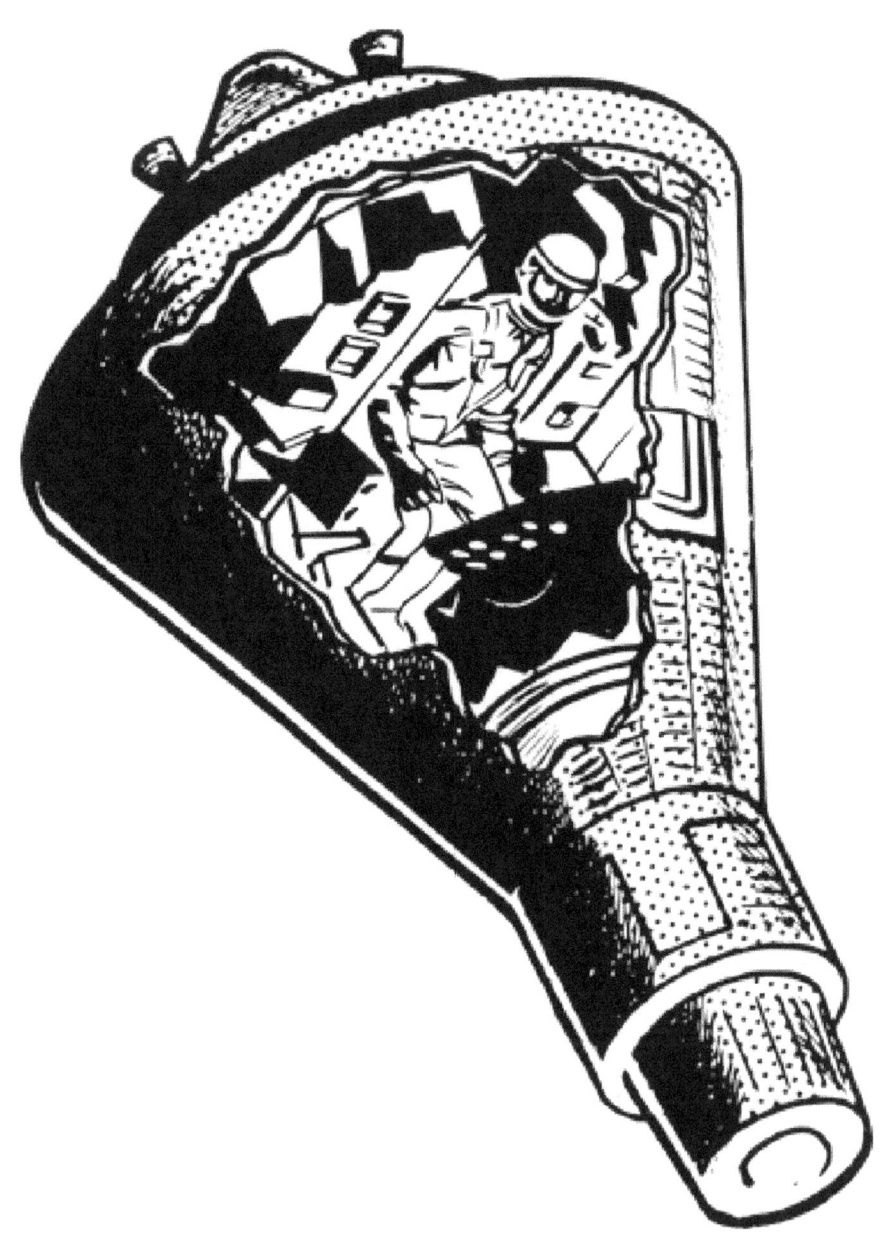

What do you think you will look like when you become and astronaut?

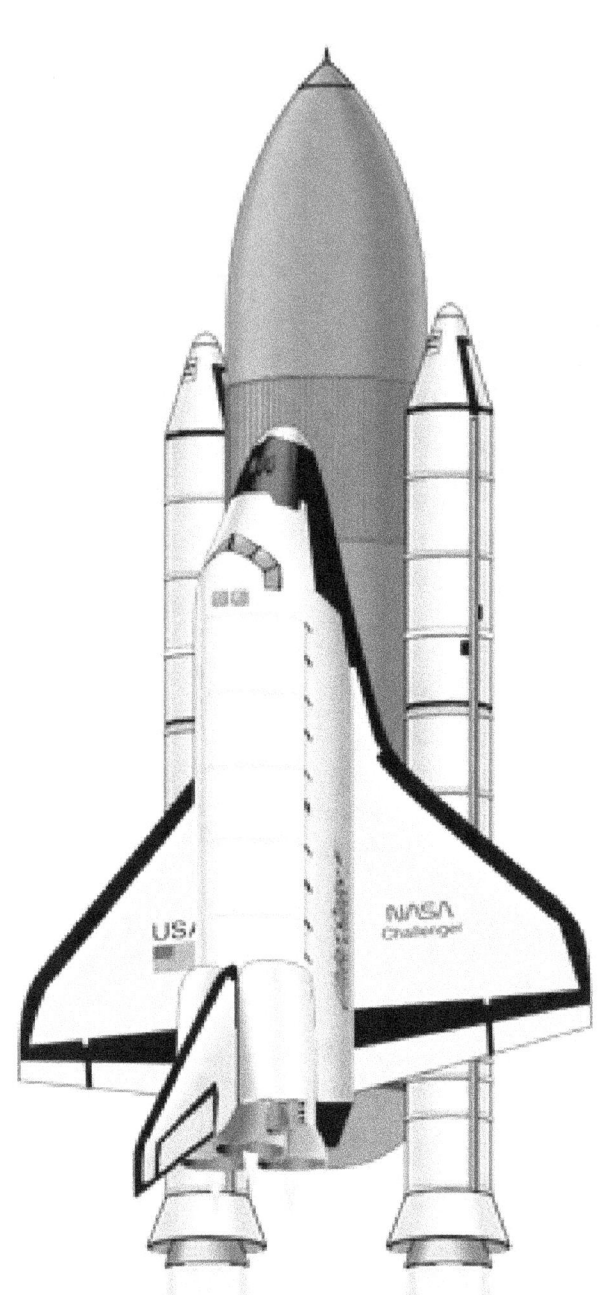

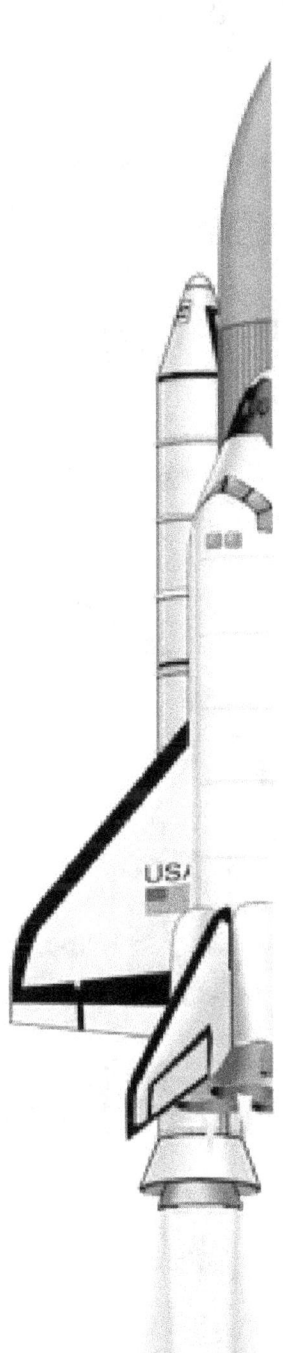

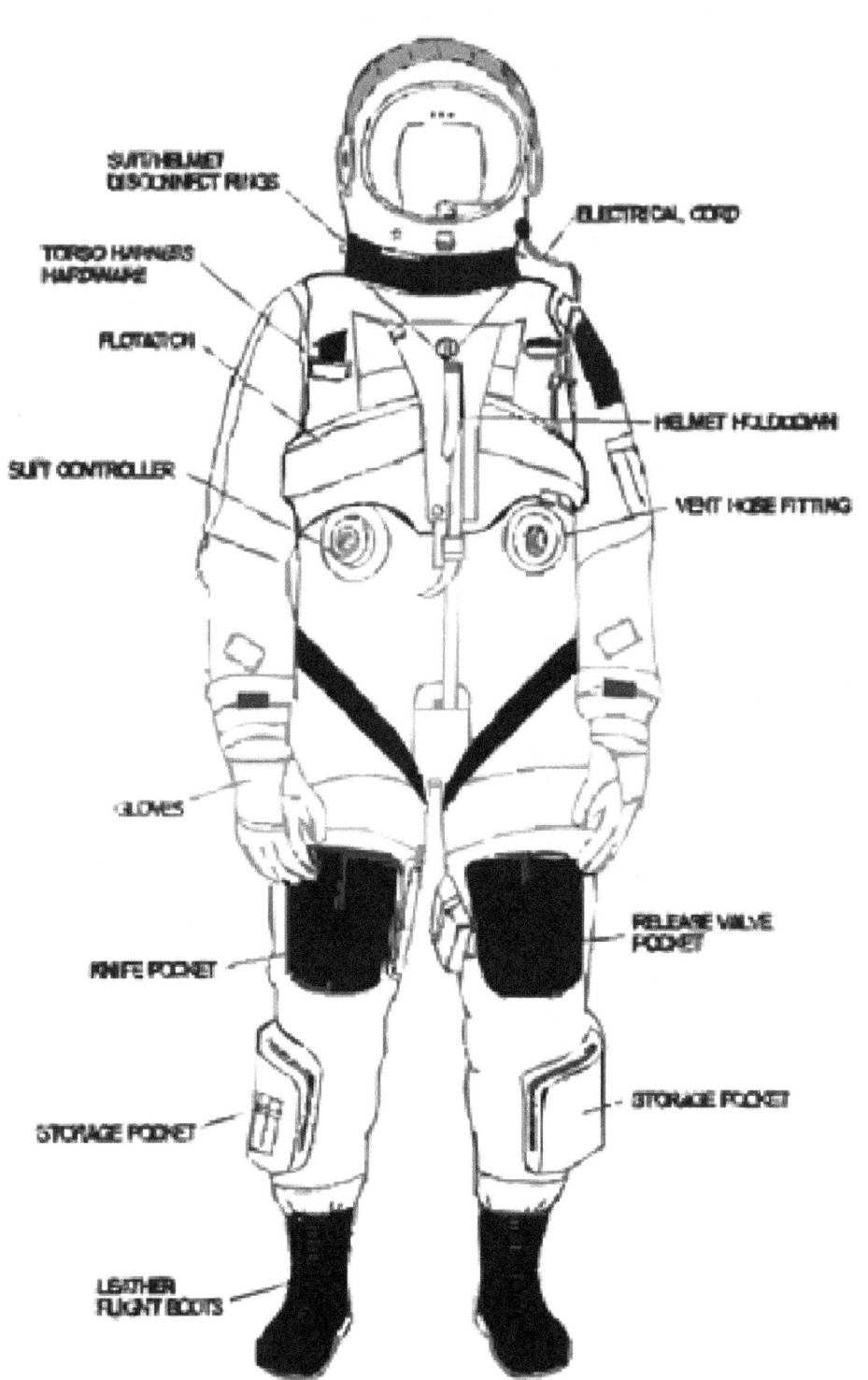

SUIT/HELMET
DISCONNECT RINGS

TORSO HARNESS
HARDWARE

FLOTATION

SUIT CONTROLLER

GLOVES

KNIFE POCKET

STORAGE POCKET

LEATHER
FLIGHT BOOTS

ELECTRICAL CORD

HELMET HOLDDOWN

VENT HOSE FITTING

RELEASE VALVE
POCKET

STORAGE POCKET

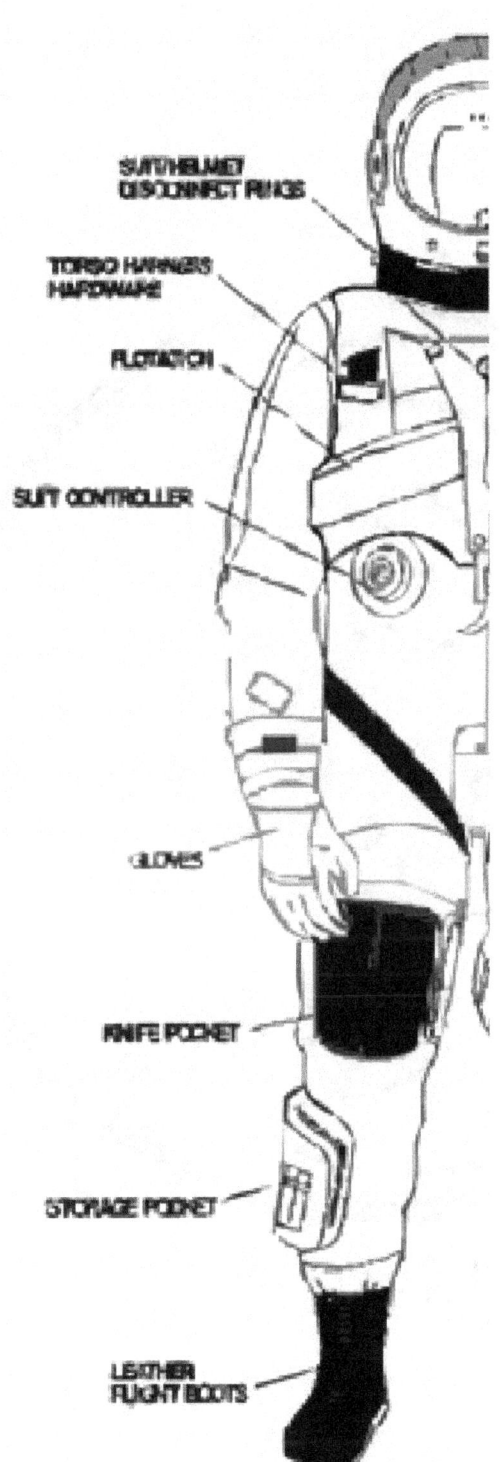

SUIT/HELMET
DISCONNECT RINGS

TORSO HARNESS
HARDWARE

FLOTATION

SUIT CONTROLLER

GLOVES

KNIFE POCKET

STORAGE POCKET

LEATHER
FLIGHT BOOTS

Mercury

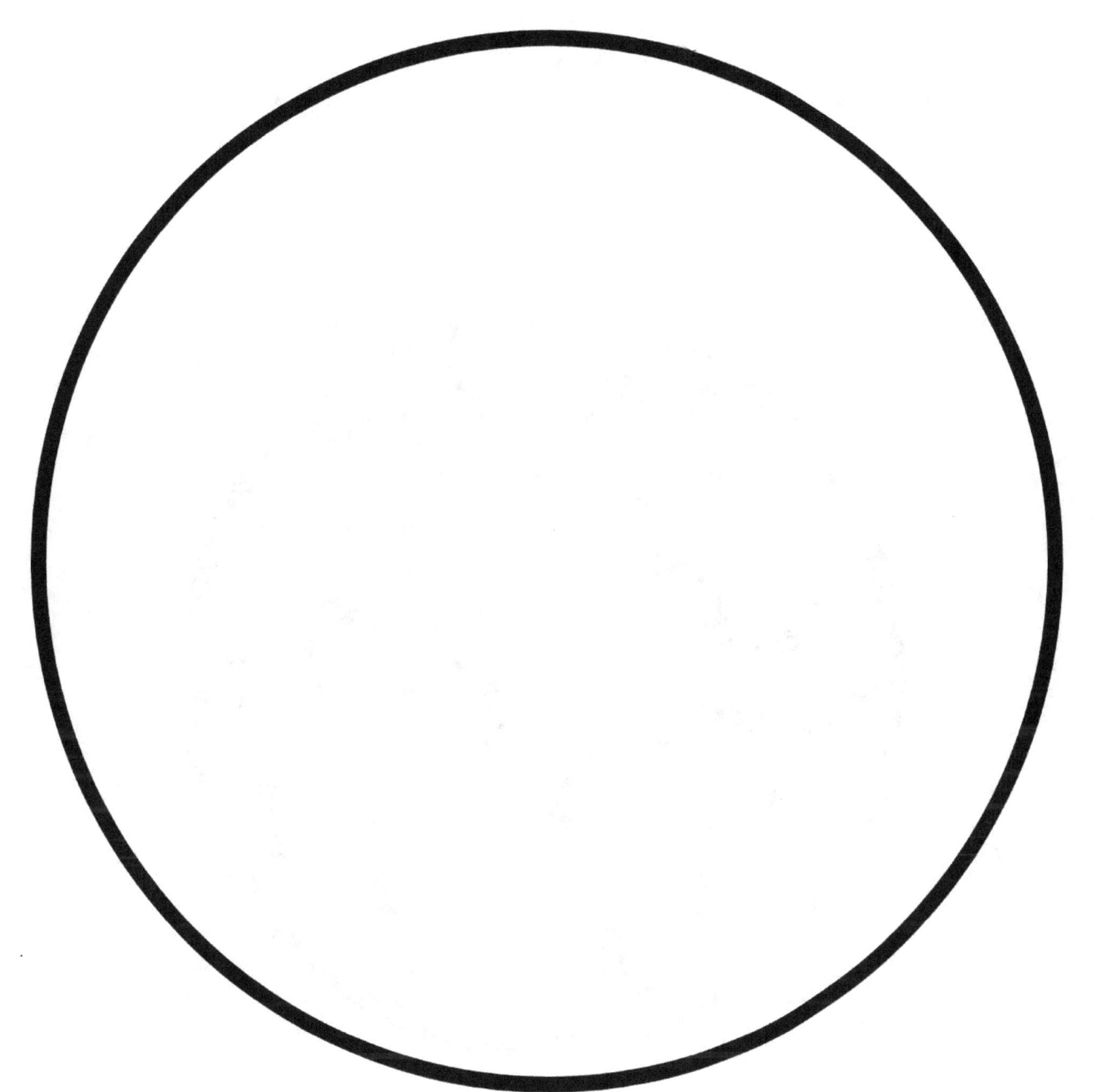

Draw Mercury.

Earth

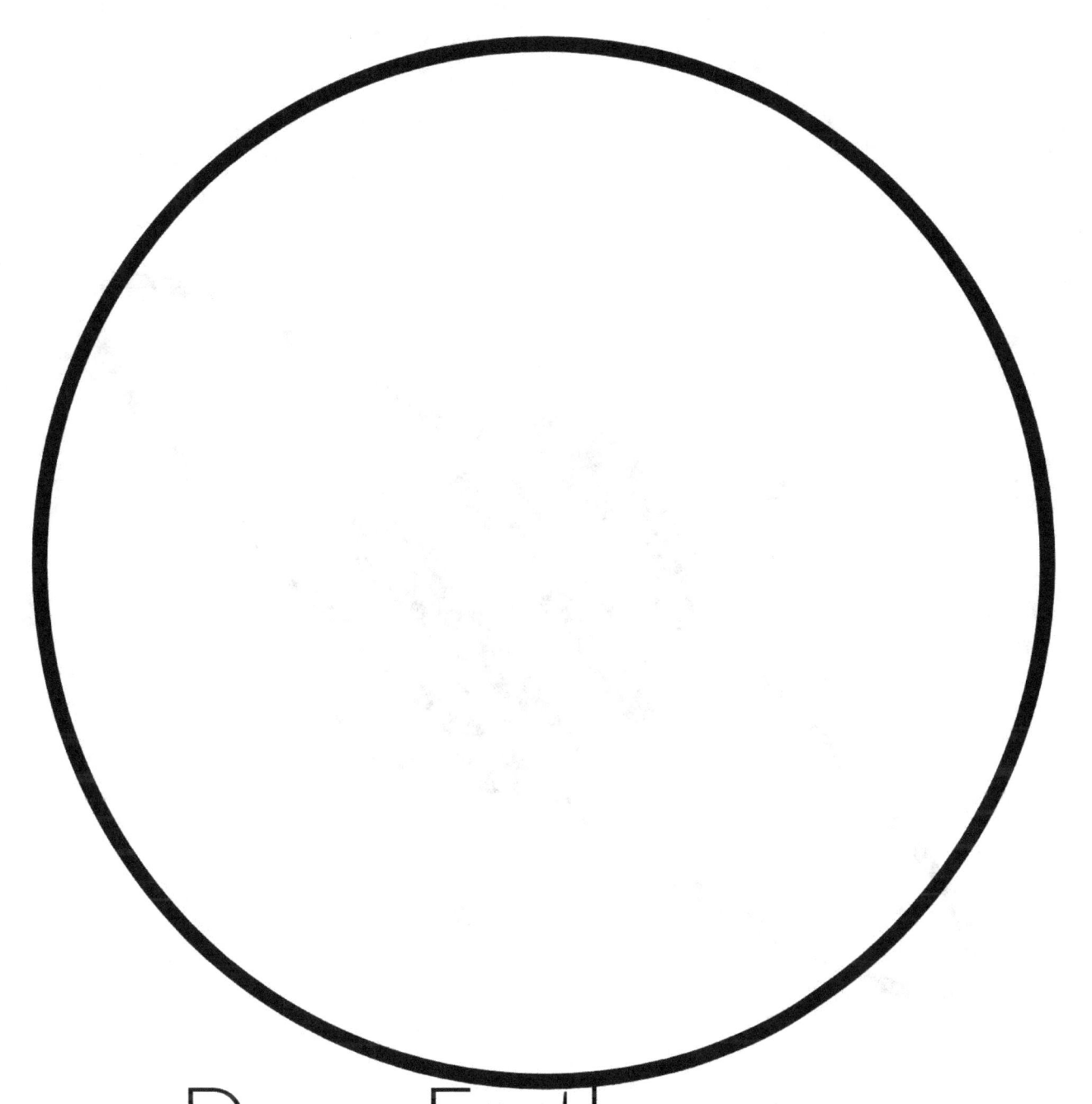

Draw Earth as you would see it from outer space.

Saturn

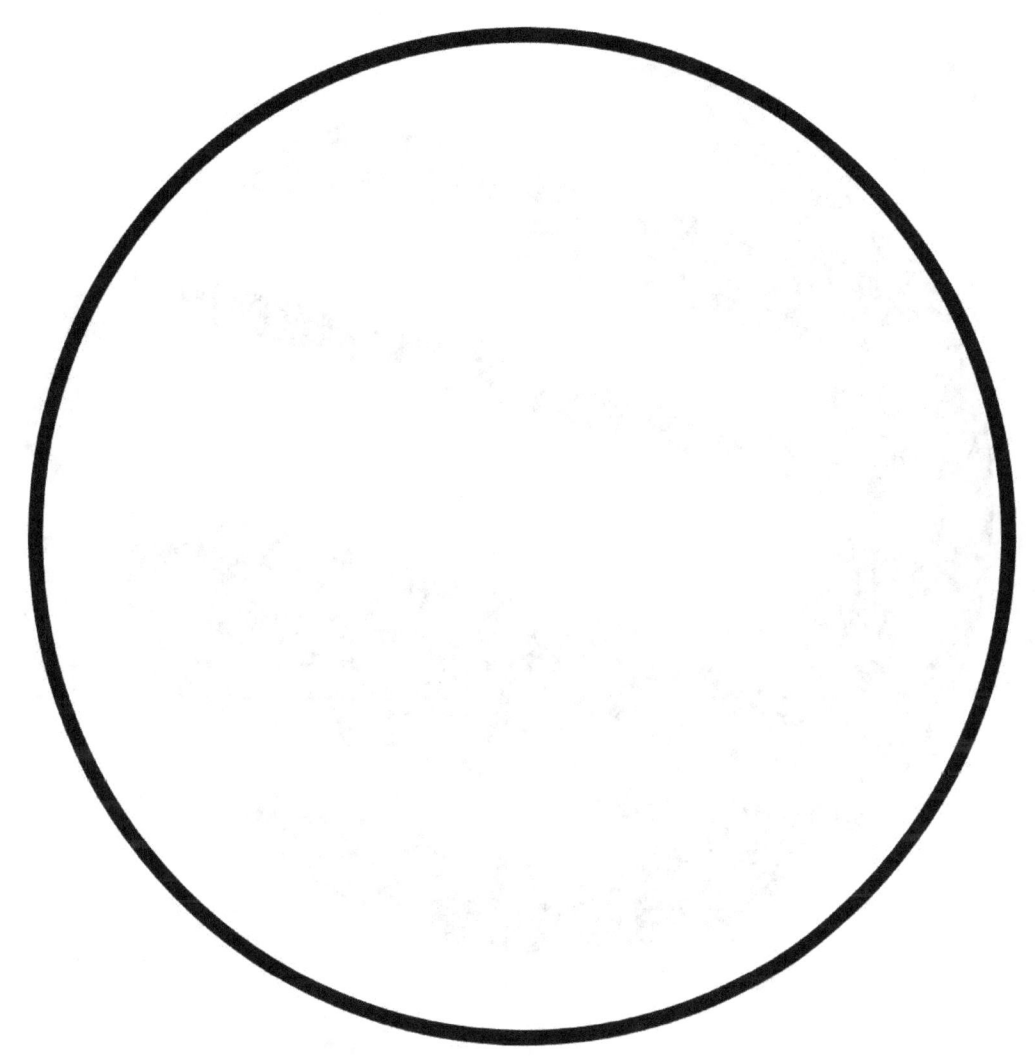

Draw Saturn.

Neptune

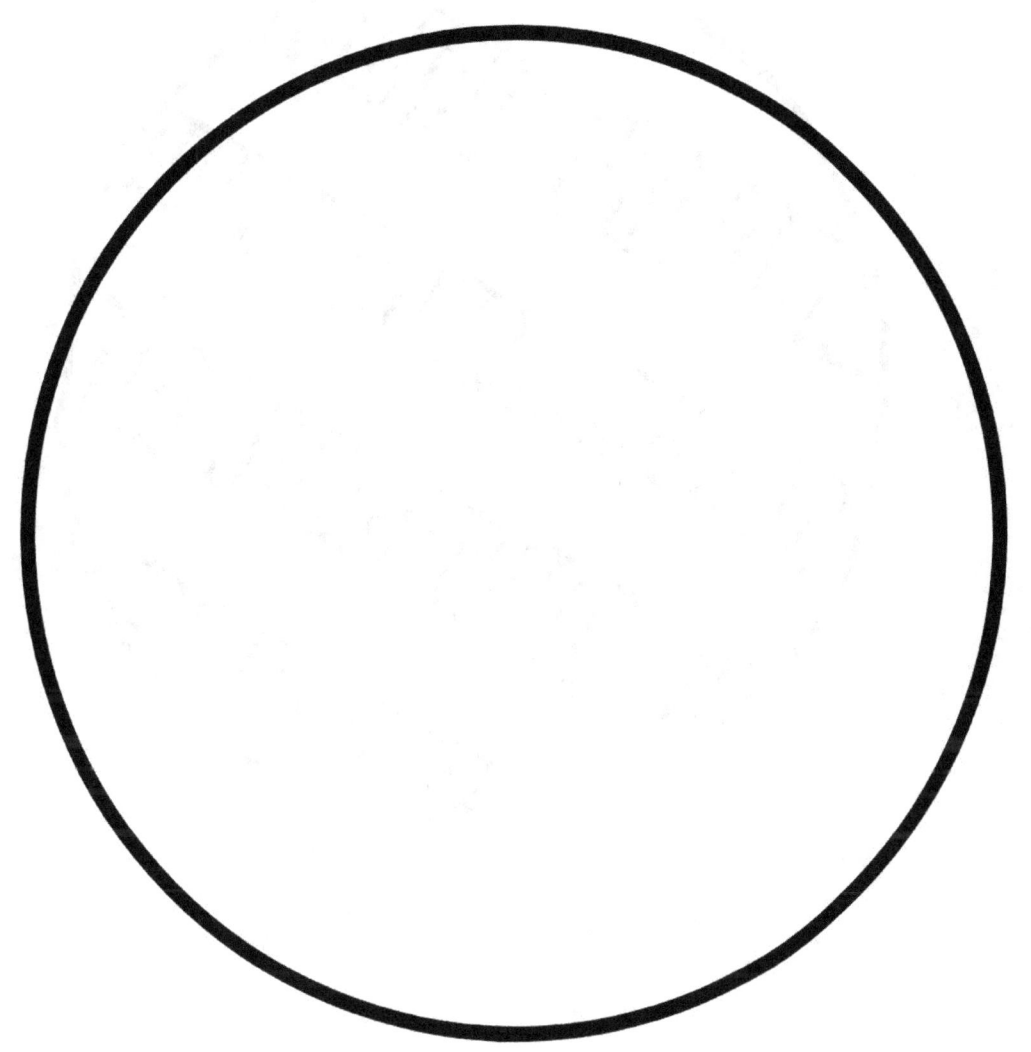

Draw Neptune.

Uranus

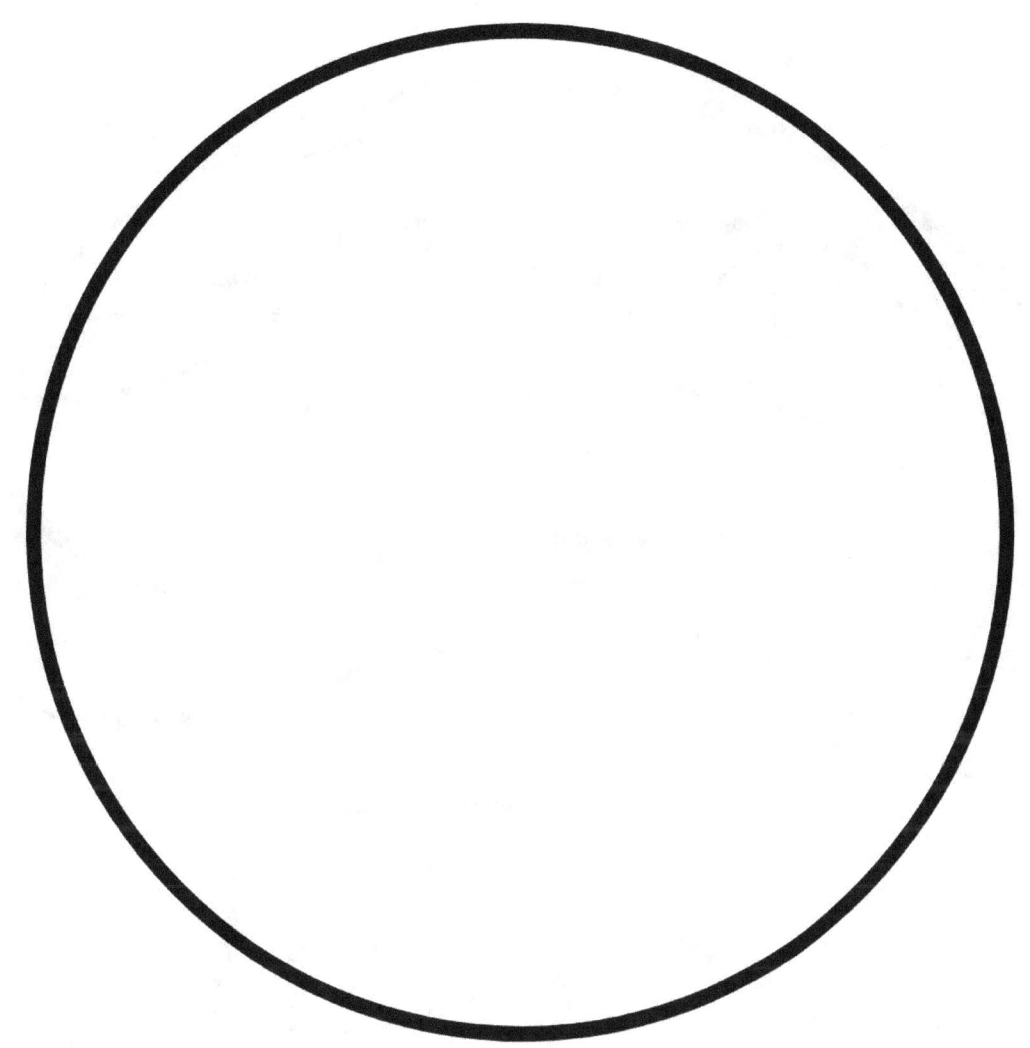

Draw Uranus.

Venus

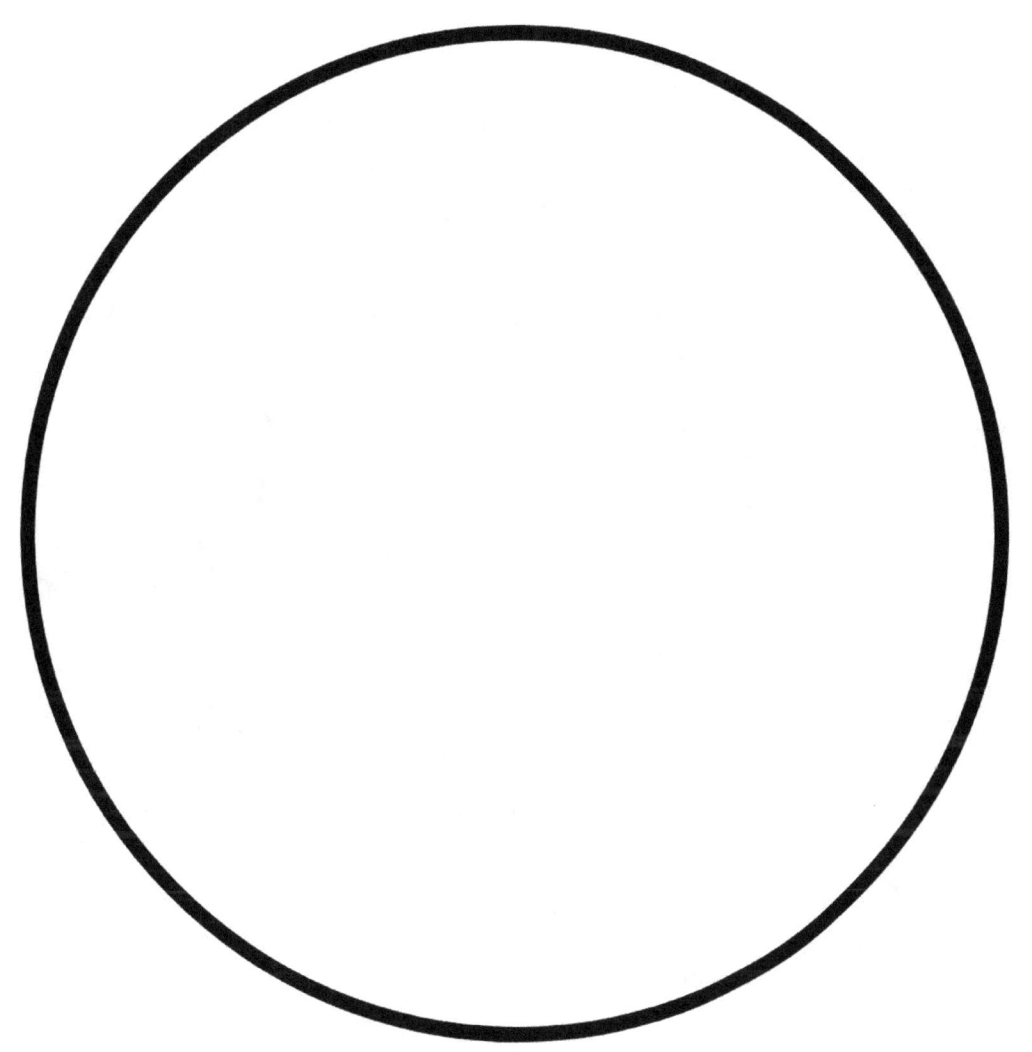

Draw Venus.

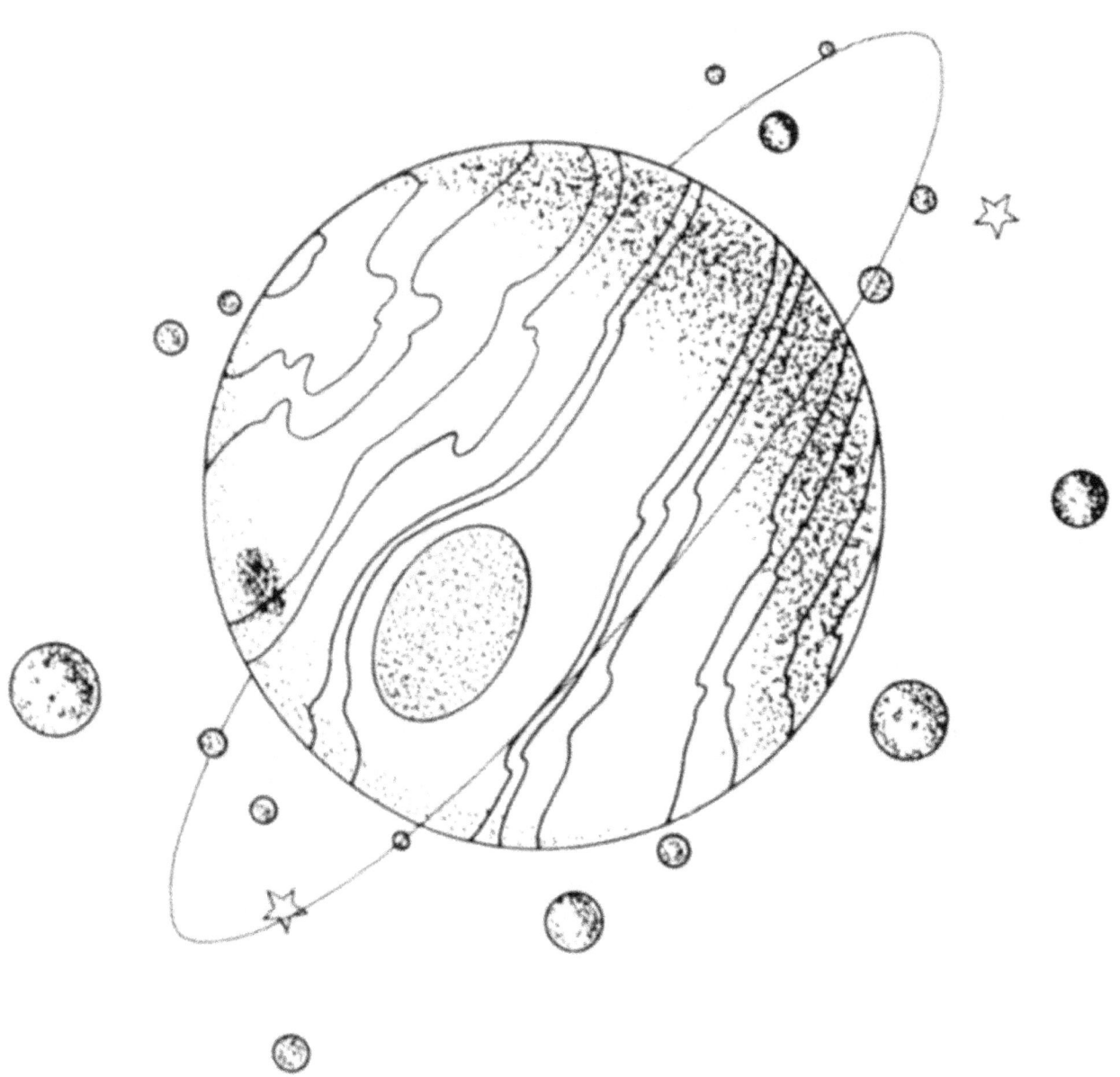

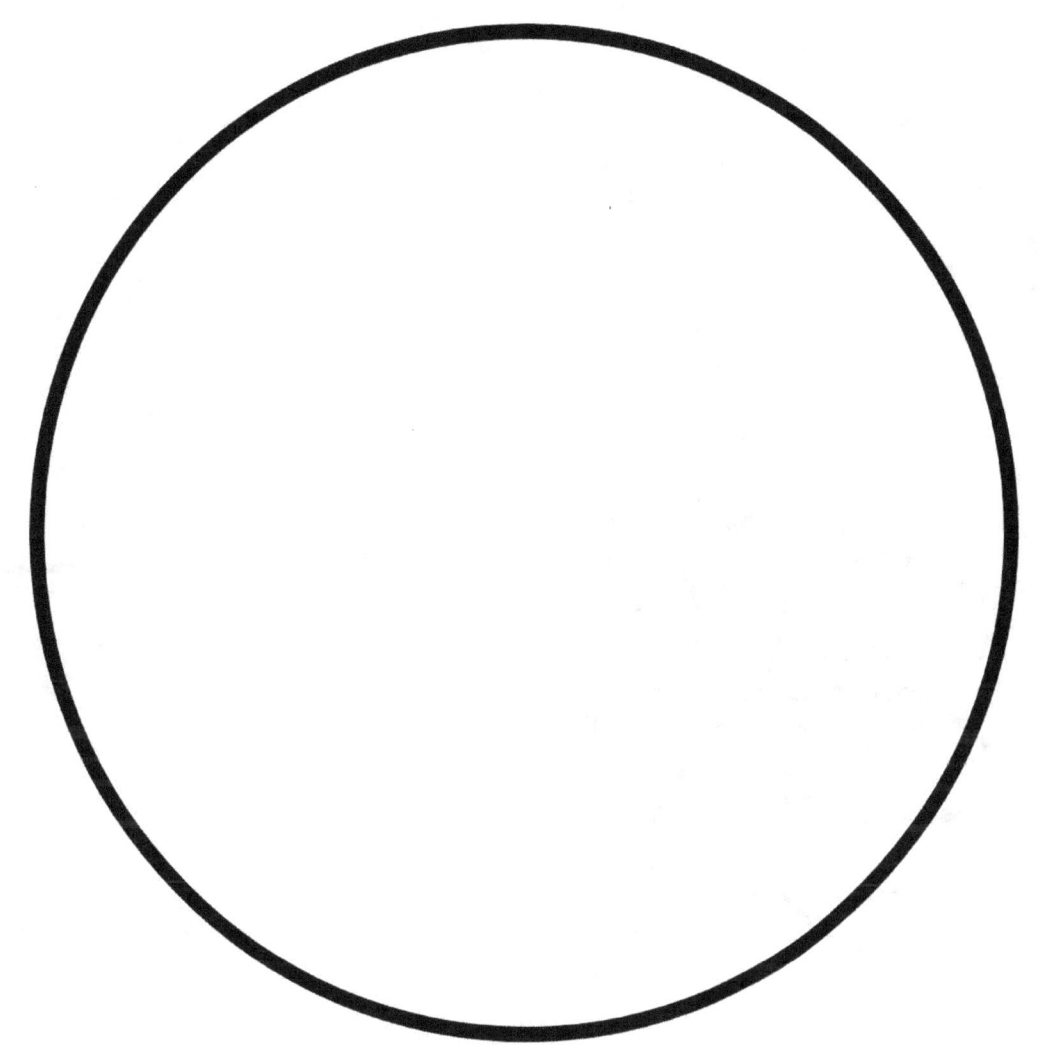

Draw your favorite
planet.

Doodle a cool moon.

Draw the missing asteroid.

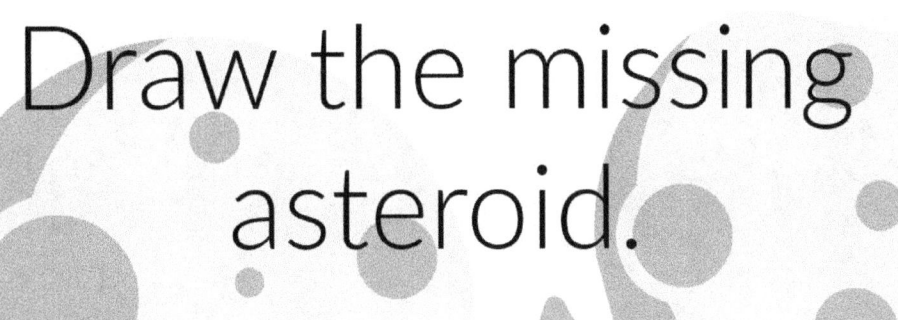

Draw the different phases of the moon.

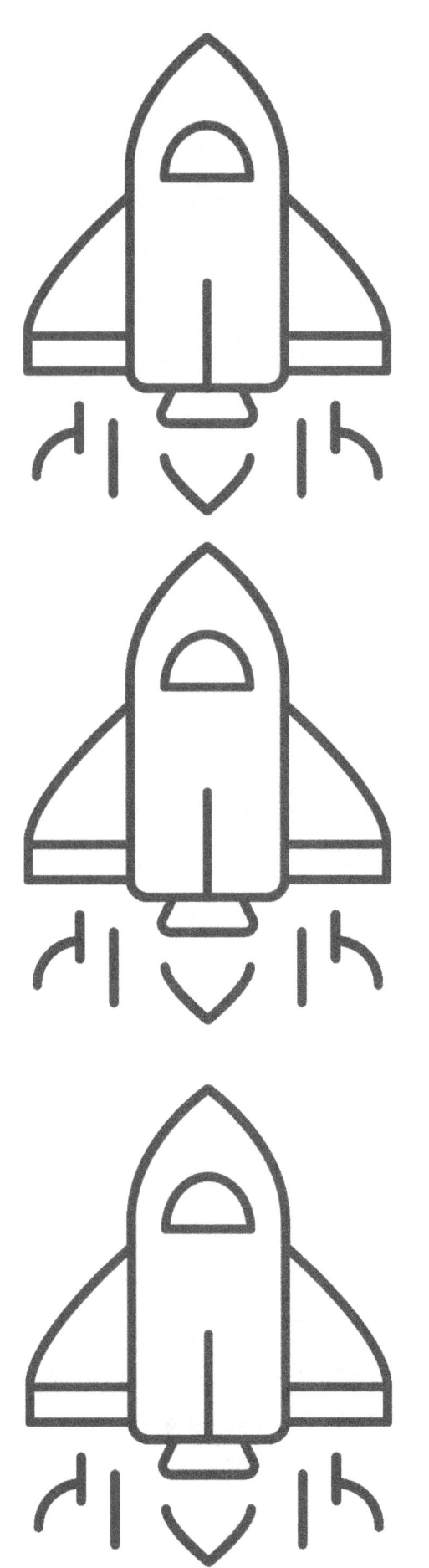
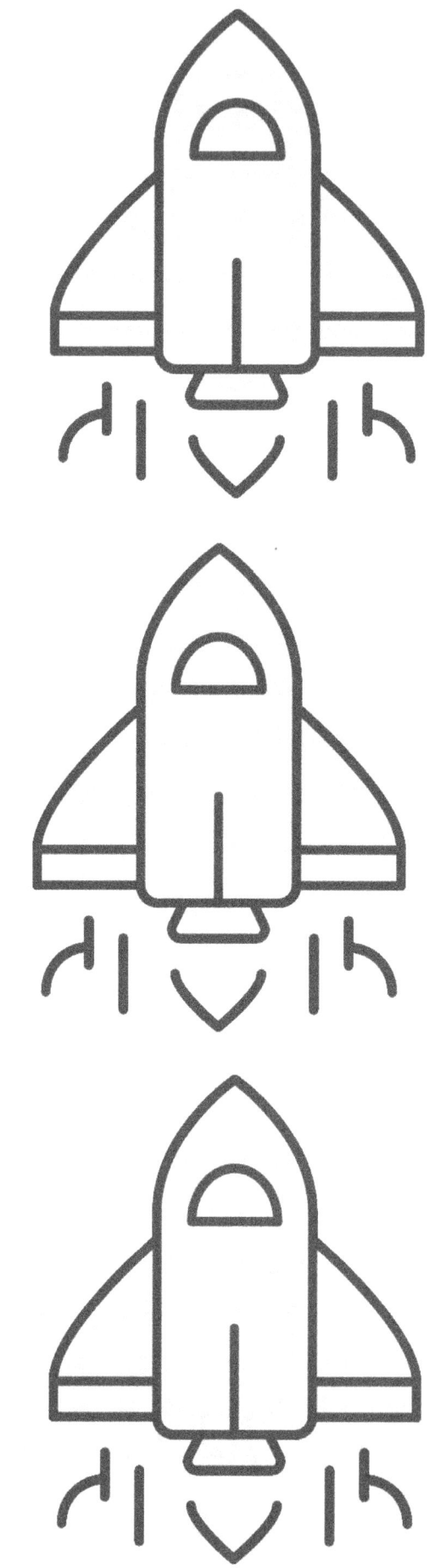

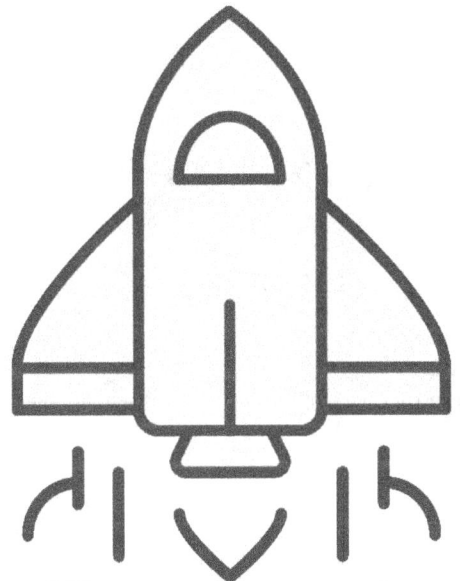

Draw the missing rockets

Can you draw the
missing suns?

What tricky line get you back to earth?

Compact
Star Burst

Draw a compact star burst.

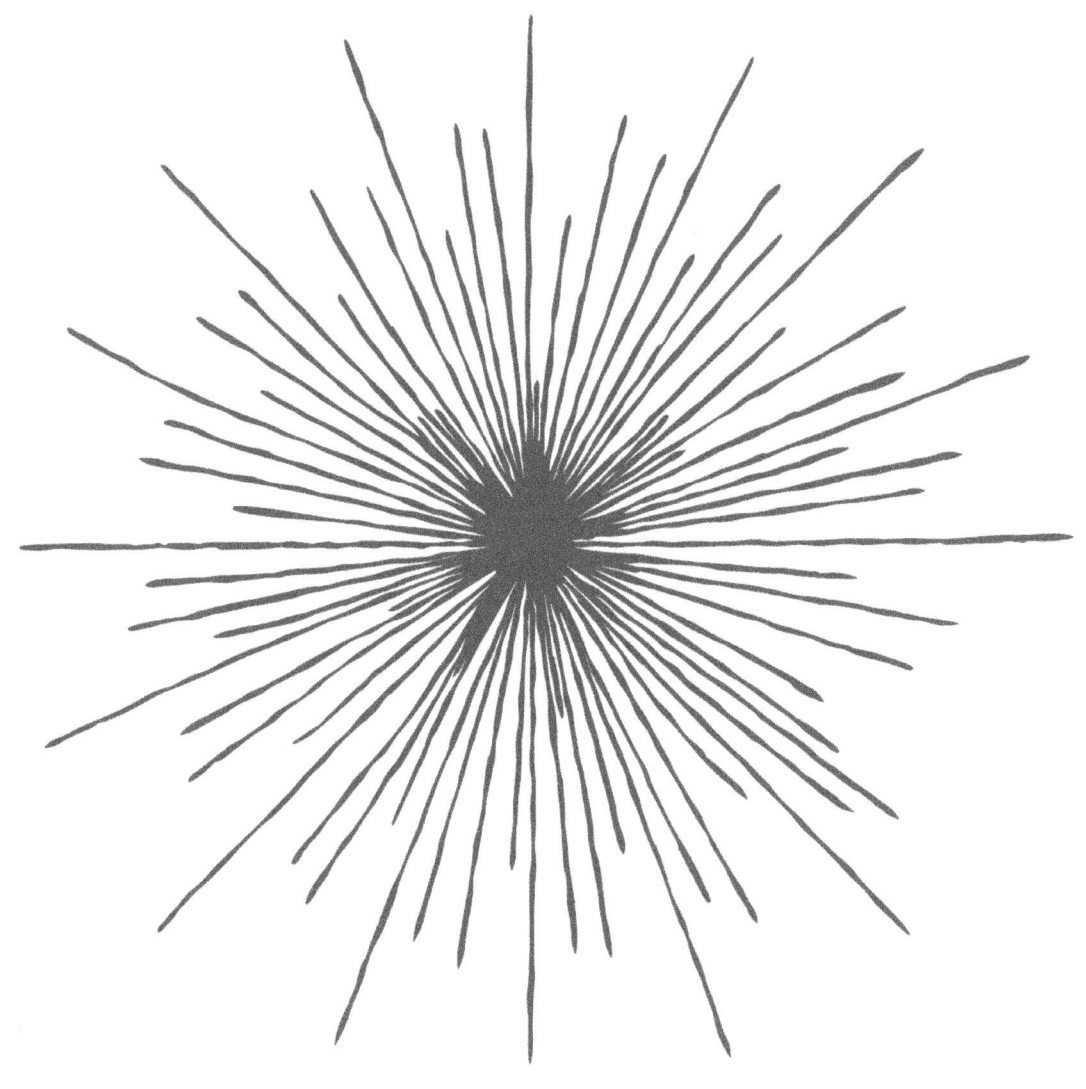

Star Burst

Draw a star burst.

Blazing Comet

Draw a blazing comet.

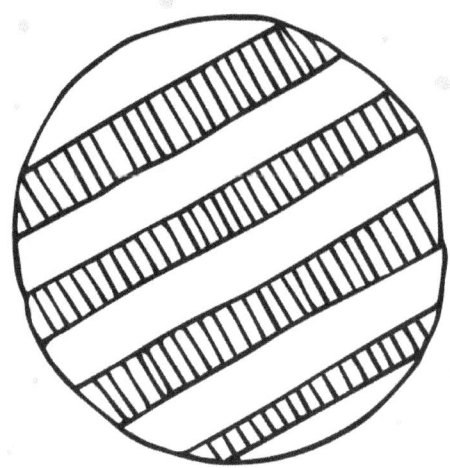

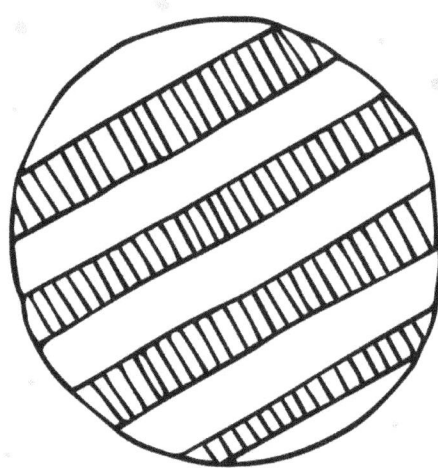
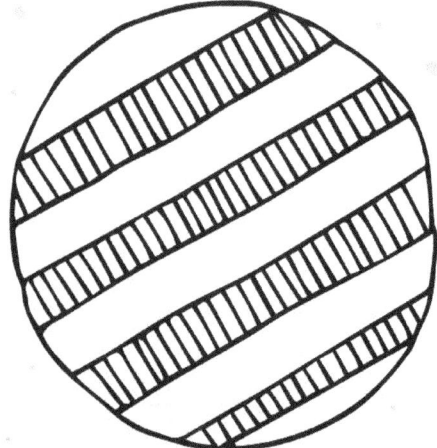

Draw a line to the twin planet.

CAN YOU FILL IN THE MISSING EARTH AND SUN?

draw in the missing moons

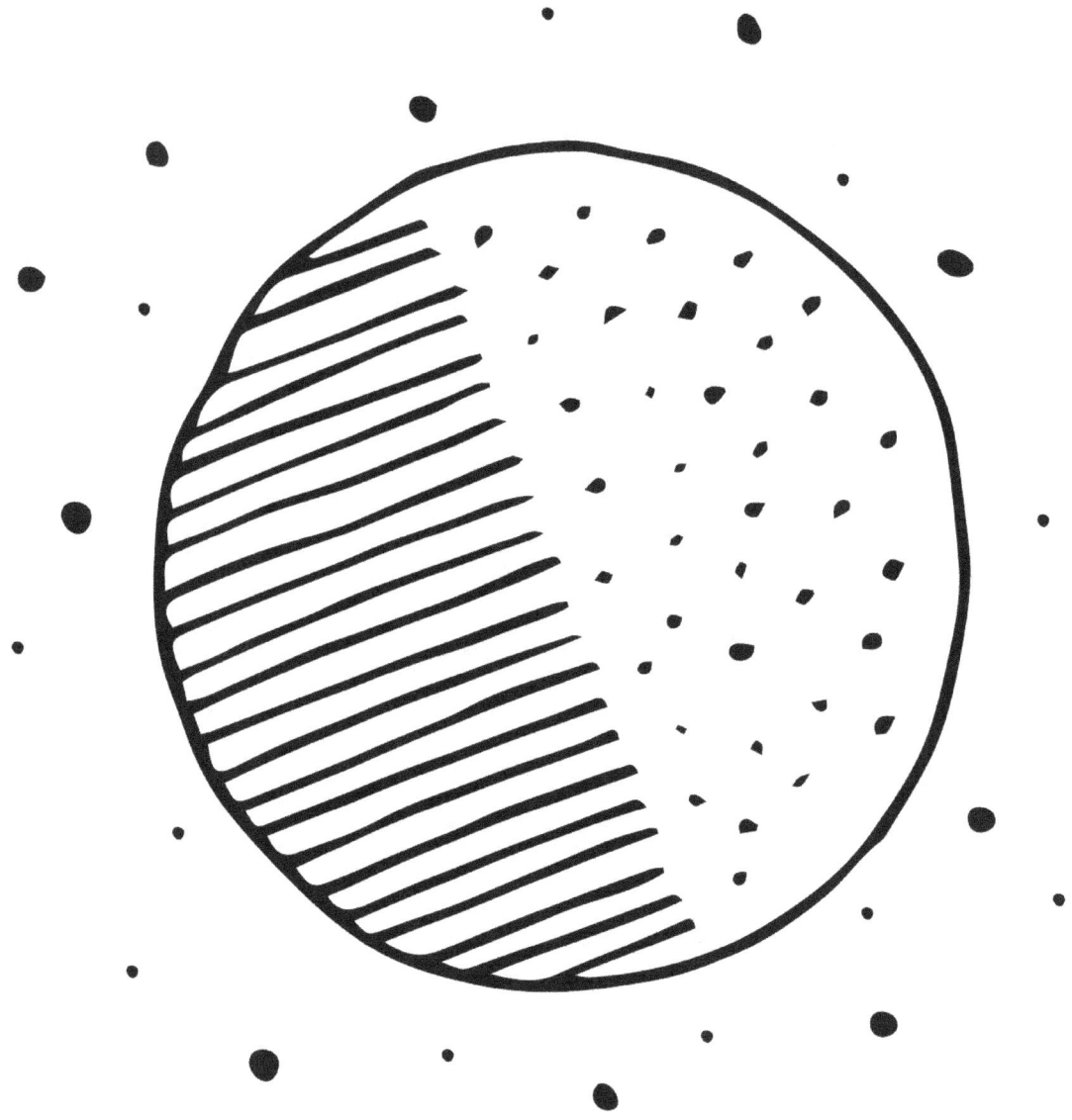

Half Moon and Stars

decorate this moon

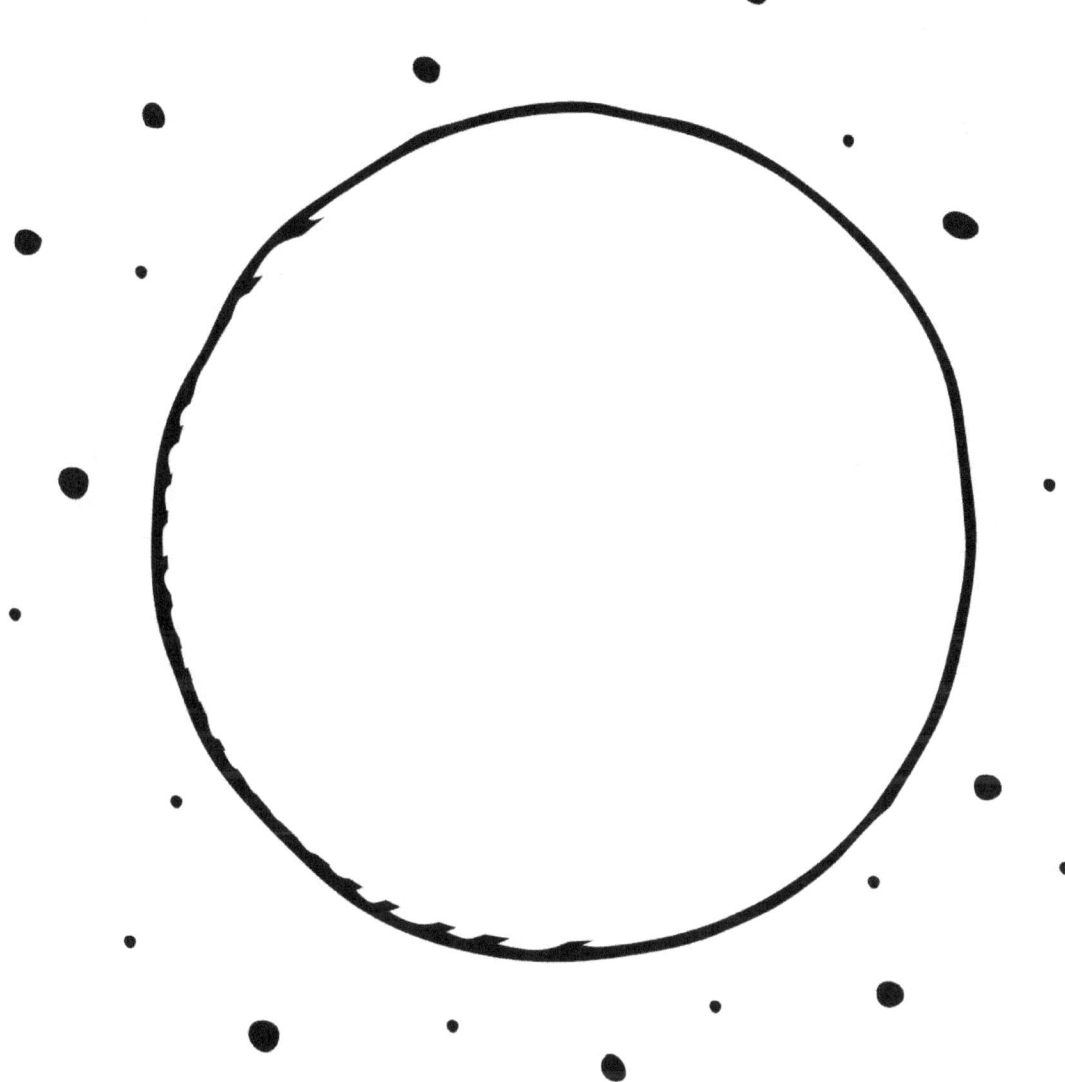

Half Moon and Stars

Wanning Moon and stars

Draw a wanning
moon.

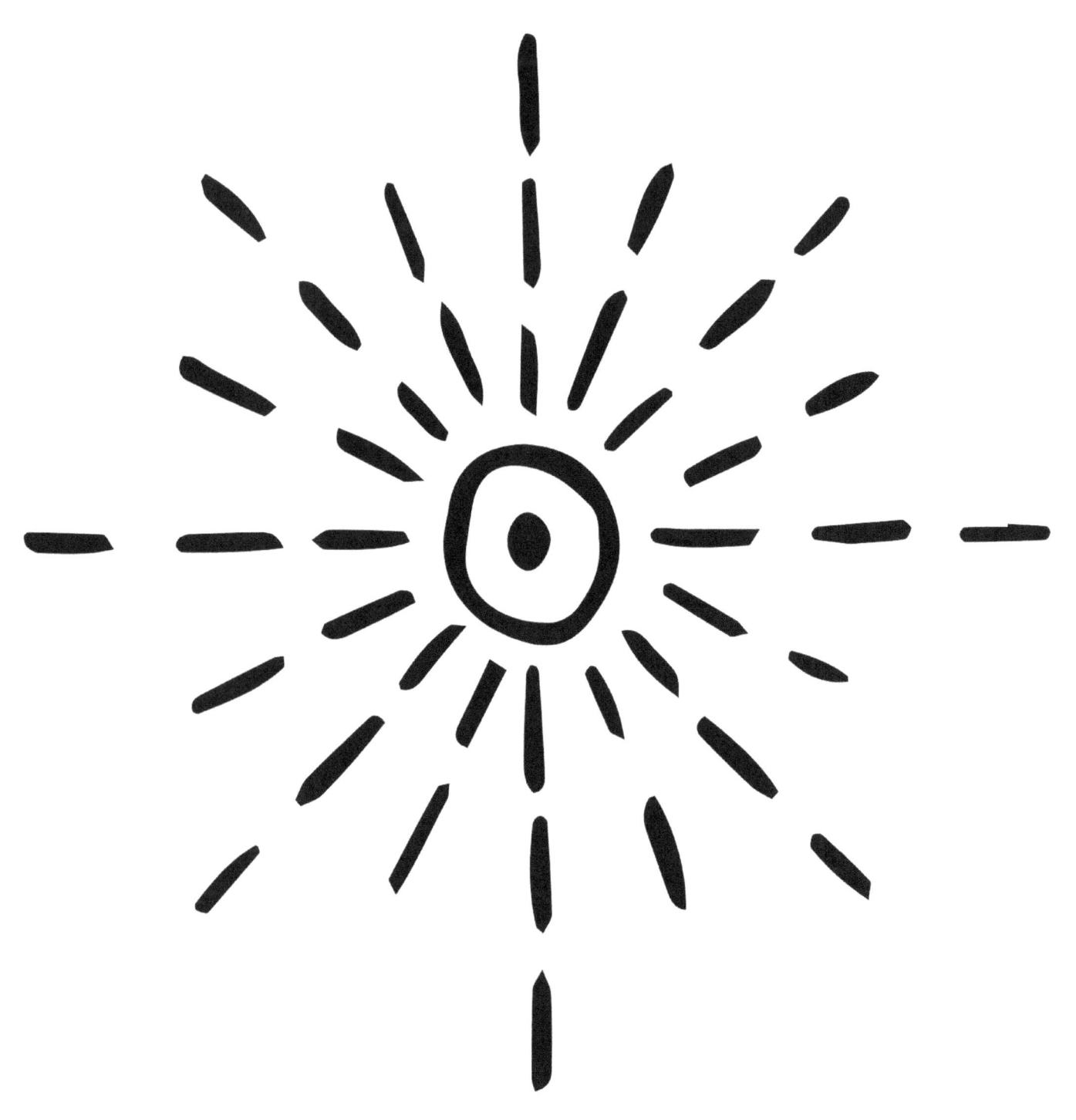

Starburst

Draw a Starburst.

Gleaming Comet

Draw a Gleaming Comet.

Crescent Moon

Draw a Cresent
Moon.

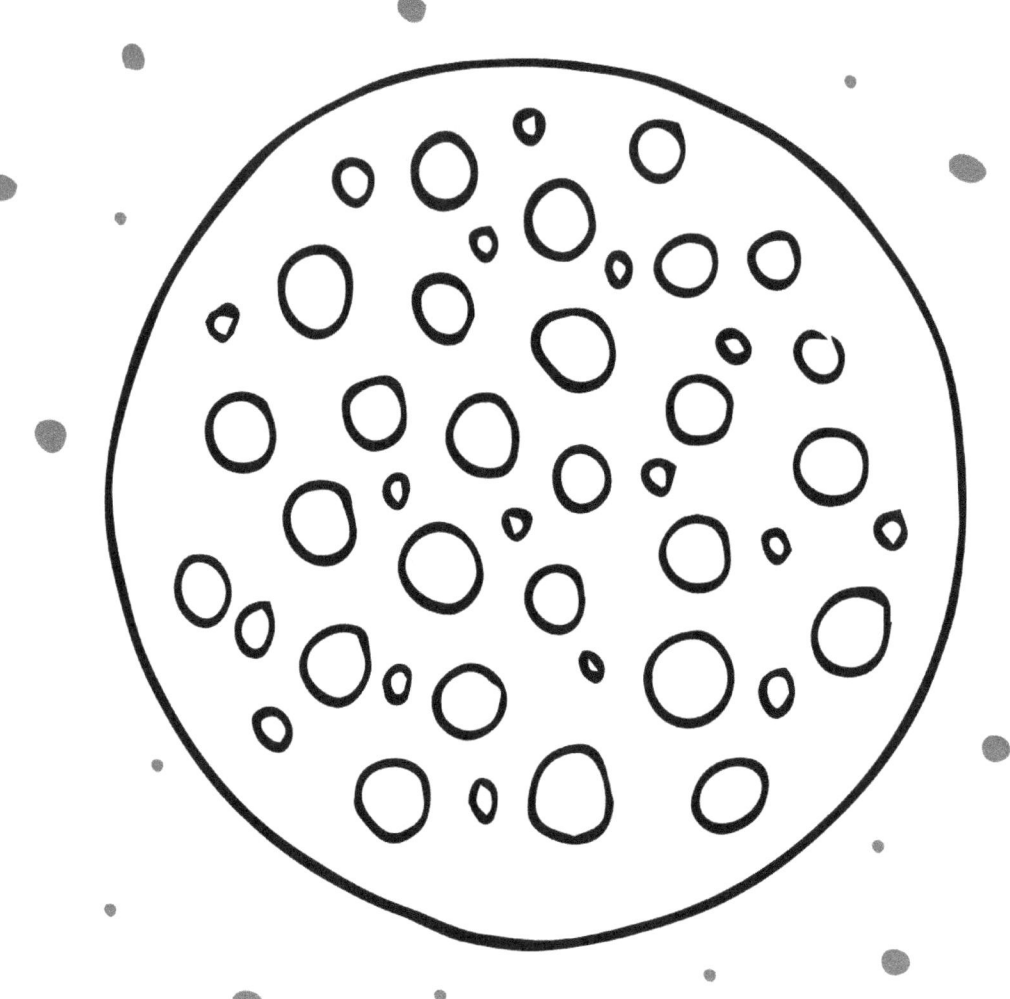

Crators of the Moon

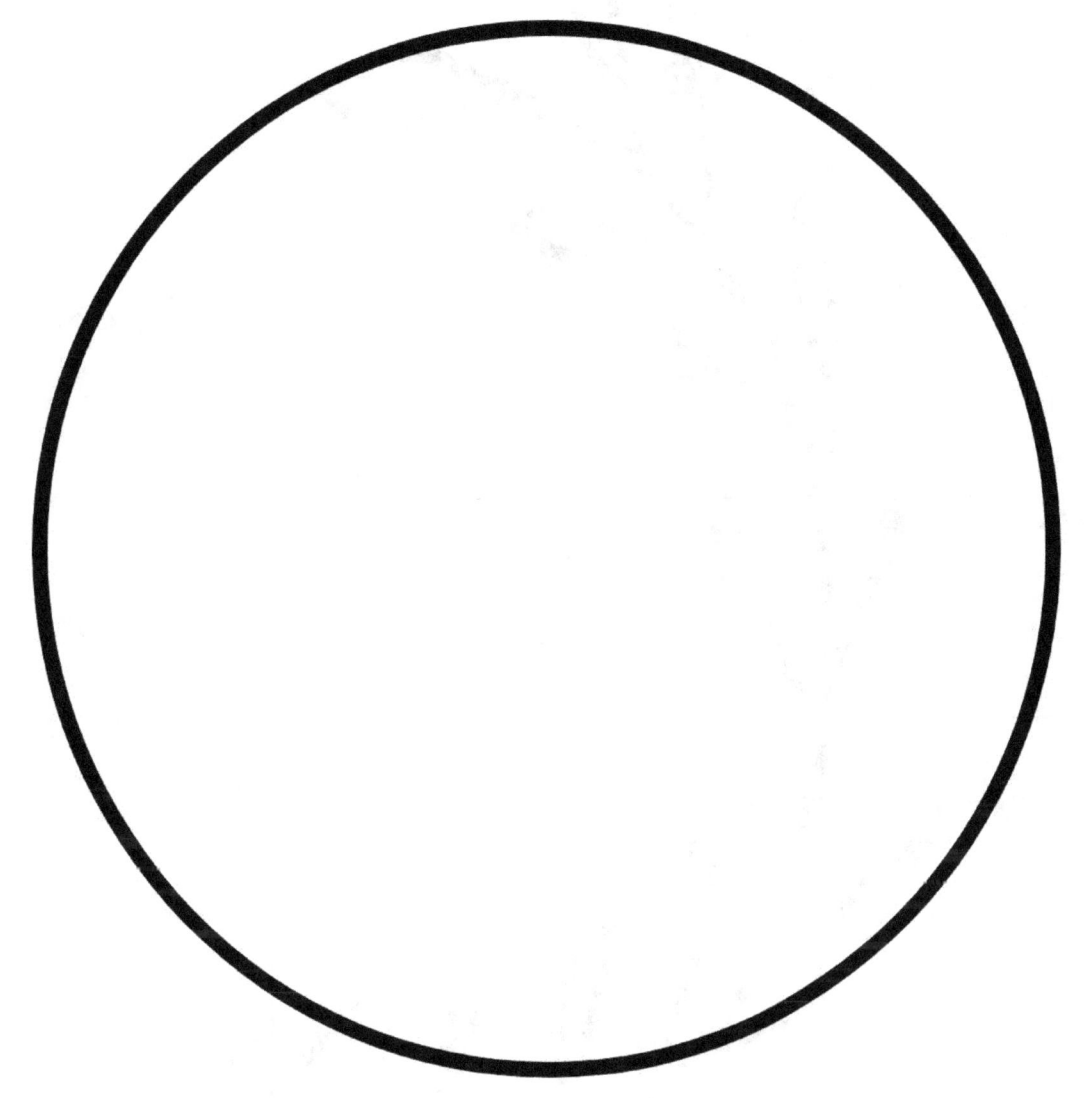

Draw craters on the Moon.

What could you draw
a moon with?

Fill in the planet's names if you can.

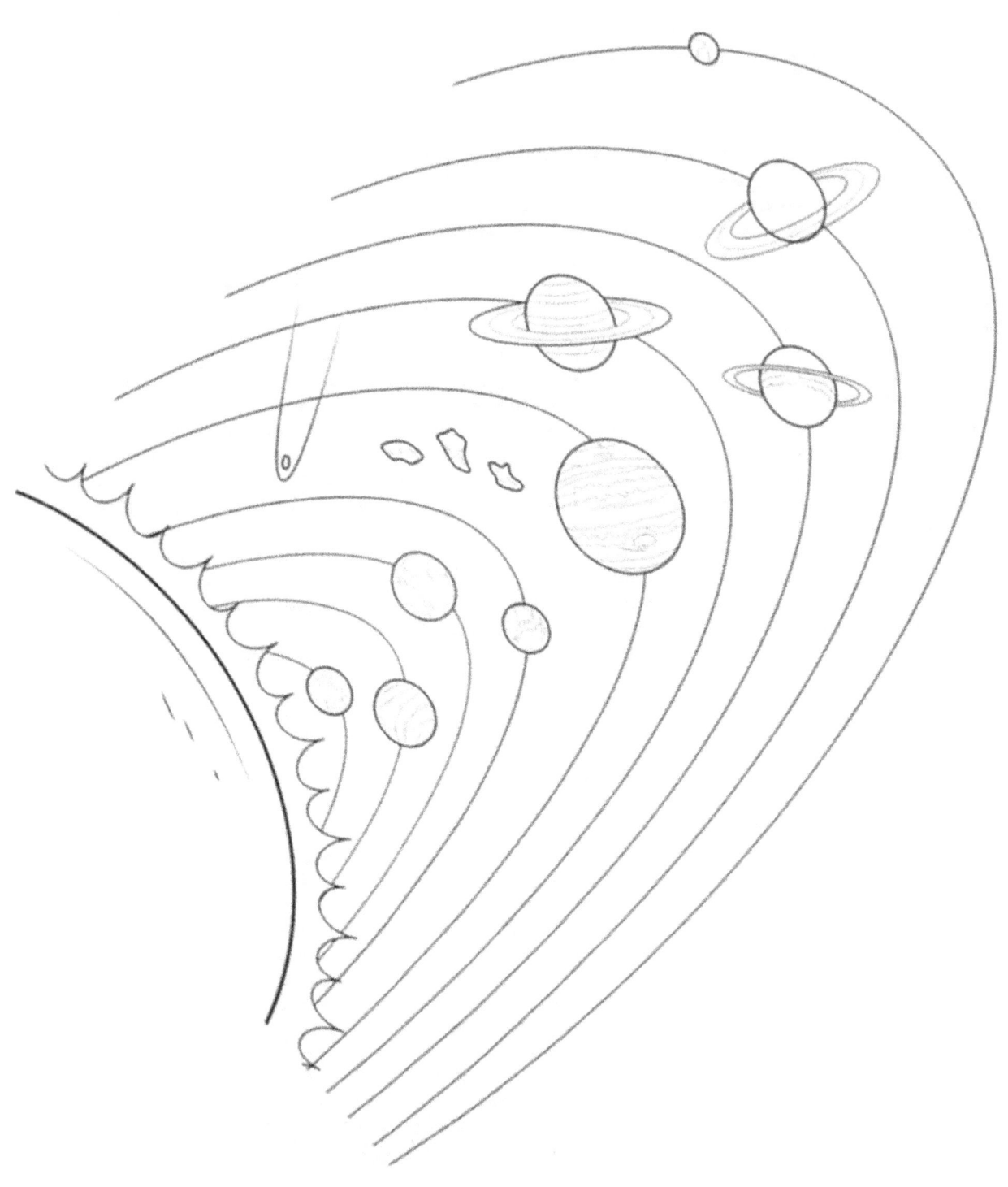

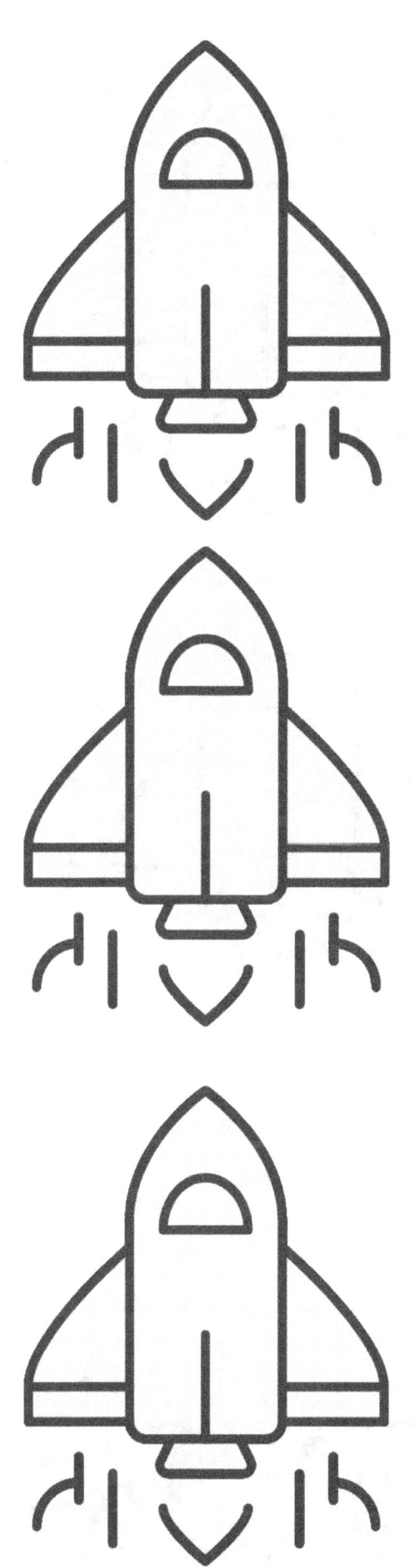
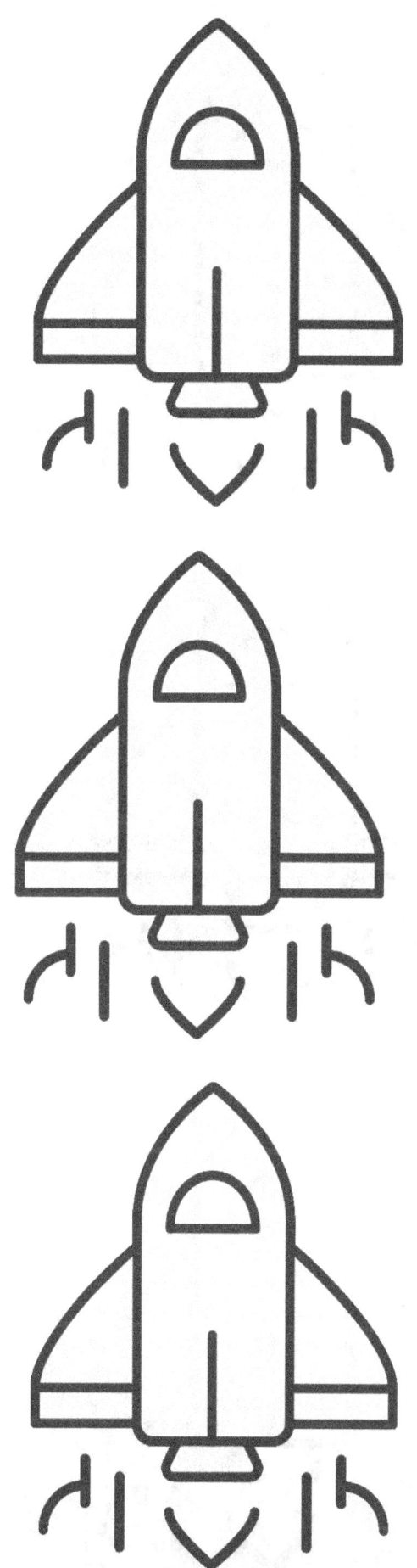

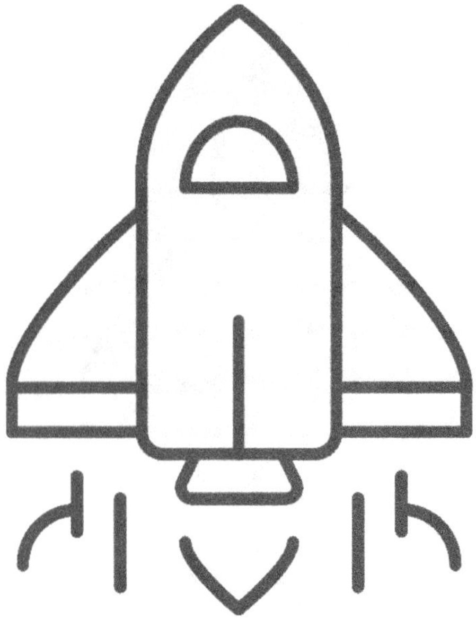

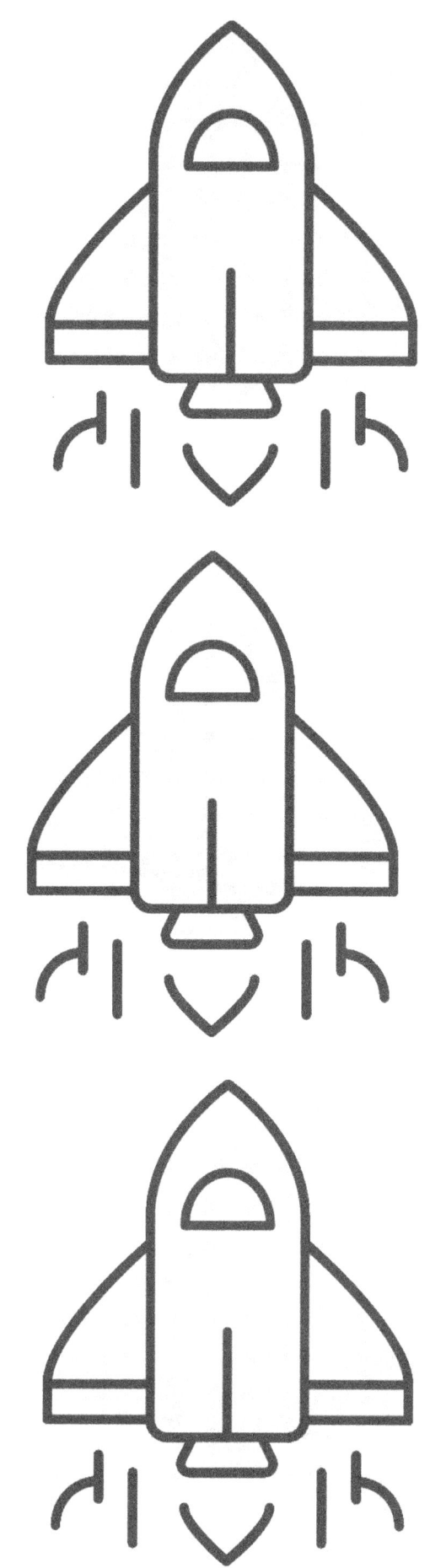

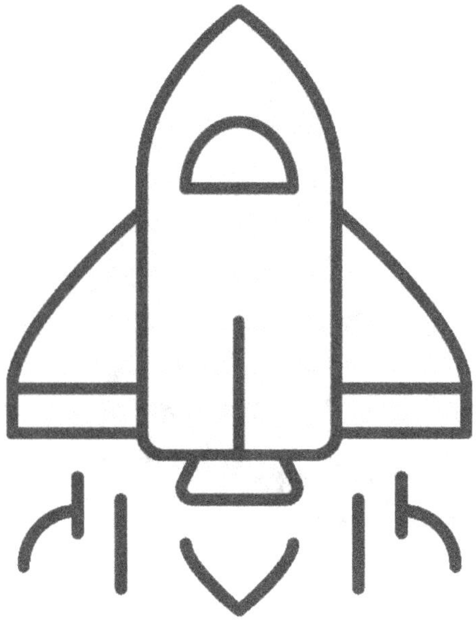

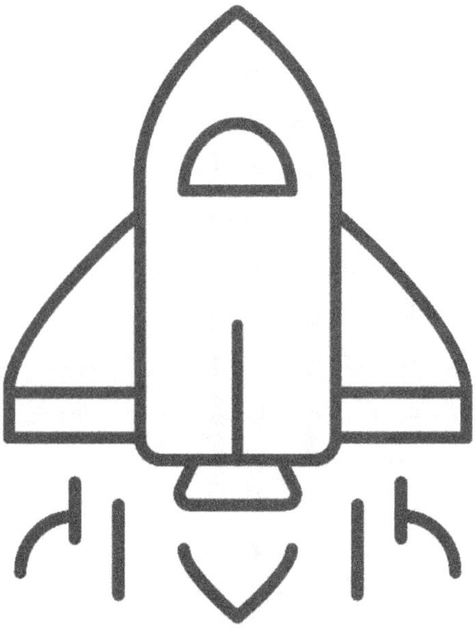